Sovereign of the Seas. On her maiden voyage around the Horn, swinging a main royal, the fore and mizzen royals having just been clewed up. Here the artist shows us better than any verbal description one of those stately white-winged racers that are no more. From painting by Charles Robert Patterson. Courtesy of Truman C. Newberry, Esq.

Donald McKay and His Famous Sailing Ships

RICHARD C. MCKAY

DOVER PUBLICATIONS, INC.
New York

Bibliographical Note

This Dover edition, first published in 1995, is an unabridged and slightly altered republication of the work originally published by G. P. Putnam's Sons, New York, in 1928 under the title *Some Famous Sailing Ships and Their Builder Donald McKay*. For the Dover edition, the original color plates have been reproduced in black and white. In addition, "Sovereign of the Seas," "Flying Cloud," "Lightning," "Westward Ho!" and "Great Republic" appear in color on the covers. Also, the plates have been repositioned in the text and descriptions of the original colors of the banners have been added to the captions for the plates facing pages 74 and 254.

Library of Congress Cataloging-in-Publication Data

McKay, Richard C. (Richard Cornelius)
    [Some famous sailing ships and their builder, Donald McKay] Donald McKay and his famous sailing ships / Richard C. McKay.
        p.   cm.
    "An unabridged and slightly altered republication of the work originally published by G.P. Putnam's Sons, New York, in 1928 under the title Some famous sailing ships and their builder, Donald McKay"—T.p. verso.
    Includes index.
    ISBN-13: 978-0-486-28820-8 (pbk.)
    ISBN-10: 0-486-28820-X (pbk.)
    1. McKay, Donald, 1810–1880. 2. Shipbuilding—United States—History—19th century. 3. Shipbuilding—Massachusetts—Boston—History—19th century. 4. Sailing ships—United States—History—19th century. 5. Naval architects—United States—Biography. I. Title.
VM140.M3M35   1995
623.8′22′092—dc20
[B]   95-10984
                                                                    CIP

Manufactured in the United States by LSC Communications
28820X05   2017
www.doverpublications.com

AUTHOR'S PREFACE

W ITH a knowledge of ships, shipbuilding and shipping affairs during the Packet and Clipper periods, gained by years of close study, and possessing original data and authentic information, I feel qualified to place before the American public a work which will be found of unusual interest and merit.

Although a comprehensive history of Donald McKay's famous ships, this work reviews America's maritime progression from the time when small packets of a few hundred tons were launched from her shipyards until she produced the largest, fleetest and finest Clipper ships afloat.

Again and again a hiatus occurs in the biographical material appertaining to Donald McKay. The master-builder of ships, completely absorbed in pursuit of his profession, accorded small heed to such things as could be preserved to an admiring posterity.

Holding that in a work of this character, the illustrative features are an essential part, I have, as the result of enthusiastic collecting for years, been able to reveal, in as far as it can be thus revealed, the wonderful development of the American sailing ship during its most progressive era. If something of the romance of those who wrought it and of the age in which they lived is also conveyed, my sublime achievement has been attained.

To Duncan McLean, friend and confidant of Donald McKay, we, of this generation, are indebted for reliable information covering the construction of ships at East Boston, which he prepared at the time for the Boston press.

I am grateful for this opportunity of acknowledging my deep and constant obligation to my Uncle, Nichols L. McKay of Boston, who loaned me certain family records, also Lieut. Commander Richard Wainwright, Superintendent of U. S. Naval Library at Washington, D. C., and Captain Frederick William Wallace of New York. Acknowledgment is now made also for the kindly advice and assistance rendered by a wide circle of friends in my search for data and some of the material contained in this book.

In requesting Mr. James A. Farrell, well-known as President of the United States Steel Corporation, to furnish a "Foreword" for this book, these personal considerations were uppermost in my mind:

He is the only American owner of a square-rigged ship sailing now under our Flag from the Atlantic Coast to the Pacific in long voyage trades. His full-rigged twenty-five hundred ton ship *Tusitala* always carries eight to ten

American boys in her crew, many of whom are now treading iron plates in prosaic craft instead of the teakwood planks of the ship in which they learned the art of sailing and navigation.

Mr. Farrell has always been genuinely interested in Donald McKay and his ships, especially the *Glory of the Seas*, which his father Captain John G. Farrell commanded and with whom, in his youth, he sailed around the Horn. At the present time he is a member of the Committee, which is going to erect a memorial to Donald McKay in Boston.

It is not given every man to have seen such wide and varied service at sea as Charles Robert Patterson, painter of ships, whose pictures, herein reproduced, form an interesting feature of this work. "Only those who brave its dangers know the mysteries of the sea," applies in truth to the marine artist, portraying conditions at sea, as well as creating on canvas the ship beautiful. With sufficient data to determine its authenticity, this sailor-painter's pictures make us see the packet and clipper ships of old when they sailed in all their glory, make us realize the beauty and grandeur of the sea, and vividly they call to our attention the fact that such ships are gone from us forever. Other well known marine artists who have contributed to the illustrative features of this work, include Fred S. Cozzens, Anton Otto Fischer, Warren Sheppard and Lars Thorsen.

For valuable assistance rendered me in various ways, I must thank Frederic de P. Foster of Tuxedo Park, N. Y., Francis B. C. Bradlee of Marblehead, Mass., Alfred S. Brownell of Providence, R. I., Miss Elizabeth Litchfield and Allan Forbes, President, State Street Trust Co., of Boston, Captain P. B. Blanchard of the N. Y. Maritime Exchange and George R. Gorman of New York.

It is claimed that the inspiriting effects of our incomparable ships in the heydey of our maritime prowess which stimulated our pride and justified our pretensions, is scarcely a memory. The present widespread interest evinced in the project to erect a fitting and proper monument to Donald McKay in Boston, proves that the ideals and traditions of a people, who, three-quarters of a century ago, led all the world in the arts and sciences of shipbuilding and seafaring, are not yet effaced.

RICHARD C. McKAY.

BROOKLYN, N. Y.

FOREWORD

HEREIN is recounted the accomplishments of a man who might well be termed the "Chief of Master Shipbuilders." Possessed of vision, a dreamer with the energy and ability to put his fancies into practical reality, a ship-designer endowed with the soul of an artist and the clear perception of an engineer, a pioneer in nautical mechanics who blazed his own path and dared when others held back, Donald McKay stands today as an epitome of those qualities which symbolized the spirit of young America. His name is indelibly connected with that high-spirited and inspiring Age of Clipper Ships—an Age which he did more than any other man to glorify with the creations of his brain and hand.

Trading ships were the rule, and freighting ships the exception when a small Rio trader named the *Courier*, sailed successfully side by side with the finest vessels afloat! This, his first production as a shipbuilder at Newburyport, Mass., brought Donald McKay before the maritime public.

A masterpiece of shipbuilding, the packet *Joshua Bates*, secured him the staunch support of New England's foremost merchant and shipowner, Enoch Train, with the result that the ambitious young shipbuilder moved to East Boston. Here he designed, as well as constructed, those splendid "White Diamond Liners" that made Train's Boston and Liverpool packets famous on both sides of the Atlantic. This builder of ships and this merchant-owner together raised New England's standard of trans-Atlantic packet service, until it reflected credit upon themselves, the City of Boston, and the country at large.

The story of Donald McKay's career unfolds a magical tale of a master-mechanic and creative artist, whose genius lifted the construction of sailing ships from the level of a journeyman's trade to the plane of an exact science. He stands pre-eminent among those acute and daring Americans whose marine achievements were the wonder of the world; made a starred banner the symbol of the earth's commerce, and enthroned the United States of America as Mistress of the Seas!

Although best known to fame as a clipper ship-designer and builder it was not only in the construction of these vessels that he excelled. Before giving to the waters his initial clipper masterpiece, he was distinguished for those speedy, strong, and well-equipped trading and packet ships—notable contributions to the art and science of shipbuilding—that rendered immeasurable service to the growth and prosperity of America and Great Britain.

FOREWORD

In 1846, Donald McKay successfully launched the *New World*, acclaimed the world's largest merchant ship, and the first three-decker ever produced. Then followed vessel after vessel, each characterized by his artistic taste and honesty in construction. The best and nothing but the best for each and every craft for which he stood sponsor!

Commerce with China and India vied with the Argonaut trade to California in bringing about the great period of the American Clipper Ship. It can be fairly claimed that a new era in naval architecture was then inaugurated. All at once, as it were, appeared a new craft of increased size, wherein speed, strength, and beauty were most happily combined—the McKay clipper, Boston-built, owned by representative Americans, and commanded by able American shipmasters! Then commenced an epoch when our maritime development placed us in the forefront among other Nations.

In his first contribution to the Cape Horn fleet—the clipper *Stag Hound*—he lengthened the model then in use. With the extreme sharpness of ends was given a large increase of the propelling power—masts, yards, canvas, etc. This craft traversed the oceanic pathway from New York to San Francisco more speedily than ever before.

It was a notable achievement, soon eclipsed, however, by the wonderful performance of McKay's next production, the famous *Flying Cloud*. She accomplished the run 'round the Horn to California's Golden Gate in eighty-nine days, twenty-one hours; and it was reserved for this same ship, three years afterward, to excel her own record by thirteen hours! Years have elapsed since then; hundreds of vessels have sailed over the same course, yet those sailing records have never been equalled.

The nations of the world were fast learning what America could do upon the ocean when Donald McKay launched his *Sovereign of the Seas*. She surpassed in size, beauty and strength all that had ever before been produced. Because no shipowner would buy so large a vessel, her intrepid builder operated her himself, and he played the merchant successfully. Although dismasted off Valparaiso, the *Sovereign of the Seas* reached San Francisco in the shortest run of the season. Homeward bound, a memorable voyage was made. Stopping at Honolulu she opened up a new mode of transportation for whale oil, etc., and carried her homeward bound pennant into port in eighty-two days. She next eclipsed all previous Atlantic crossings and beat the Cunard Steamship *Canada* en route to Liverpool. Leaving Sandy Hook on the same day that the Cunarder left Boston the *Sovereign of the Seas* arrived off the bar at Liverpool when the *Canada* arrived at Queenstown. Ere this smart clipper came to her end she made some remarkable passages in the Australian and other trades.

Throughout the most brilliant period in the annals of American shipbuilding, Donald McKay maintained the lead. Ships of surpassing beauty and increased size, showing superiority in design and workmanship, were launched from his yard at East Boston. Within a remarkably brief period the *Staffordshire*, *Flying Fish*, *Westward Ho*, *Chariot of Fame*, *Empress of the Sea*, *Romance of the Sea*, and other wonderful models of clipper architecture, were dashing through to the equator, encountering baffling winds to the southward and battling with gales 'round the Horn, and beating along South America's Pacific Coast en route to California's El Dorado.

The improvement of sailing vessels continued and reached a climax when McKay's clipper *Great Republic*, the largest ship in the world, was completed in October, 1853. He designed this, his masterpiece, to be the fleetest of his craft. Daringly, with supreme confidence in his own work and the future of American commercial enterprise, he built her entirely on his own account. Fate ordained that his hopes should be shattered, for she was burned to the water's edge before embarking upon her maiden voyage. The *Great Republic* was truly a wonderful ship, for her performances after rebuilding and even under reduced rig left little room for doubt that she would have proved all that her builder claimed.

Now came England with her interest in Australia,

and engaged Donald McKay to build the famous Clipper Ship Quartette—*Lightning, James Baines, Champion of the Seas,* and *Donald McKay.* This is the only time in history that an American shipbuilder not only laid the keels of ships to fly the English flag but built a fleet that shed lustre upon Britain's ensign.

The *Lightning* upon her maiden voyage made the phenomenal run of four hundred and thirty-six miles in twenty-four hours, which entitles her to the proud distinction of being the swiftest ship that ever sailed the sea! It was thirty-five years before an ocean-going steamship exceeded that day's work. Her remarkable passage from Melbourne to Liverpool in sixty-three days has never been equalled under sail.

A record-breaking run, Boston Light to Rock Light, Liverpool, in twelve days, six hours, brought the *James Baines* prominently to the fore. This marvellous ship made the quickest passage, Liverpool to Melbourne, of sixty-three days, eighteen hours, returning home in sixty-nine and one-half days, thus sailing 'round the world in one hundred and thirty-three days. This entry in her log on another voyage, *"going 21 knots with main-skysail set,"* is the highest authenticated sailing record.

Upon her maiden voyage the *Donald McKay* distinguished herself by sailing from Boston to Cape Clear, Ireland, in twelve days, recording, en route, a twenty-four hour run of four hundred and twenty-one miles.

For many years she was the largest sailing vessel afloat.

Such remarkable contributions to Great Britain's shipping had a worldwide effect. Today it must be recalled that besides proving the supremacy of American shipbuilding, McKay's clippers and other American-built sailing vessels, took a prominent part in forging the link which connected England and Australia.

As a modification of the swift-sailing "extreme" clippers came the half—or medium—clipper, ships with less spread of canvas wherein cargo-carrying capacity came before speed. Here again Donald McKay took the lead. He constructed many of this type, his last being the *Glory of the Sea*, which stand favorable comparison with sailing craft of all times. The decadence of American shipbuilding, traceable to a variety of causes, commenced about the same time, and Donald McKay, after a long, long struggle, closed his shipyard.

Donald McKay was the presiding genius of the designer-builders who brought the sailing ship to the acme of perfection at a period when sail was engaging in a battle for life with steam. Under the lash of competition, he put forth his best efforts and produced sailing vessels which, for a time, almost checked the advance of iron hulls and steam propulsion. But it was, after all, a losing fight, and, no doubt, in his heart he knew it. Yet his was the soul of a fighter, a man whose best efforts were called up in the stress of competition. And the pace he set was, perhaps, a spur to the opposition, causing them to redouble their energies in perfecting craft that would break man's dependence upon the wind and the disappearing forests.

Men have since designed and constructed vast and wonderful fabrics for seafaring, but none of these master-minds in the fields of mechanics and engineering loom so large in the records of shipbuilding as that of the East Boston shipbuilder, Donald McKay. His genius, his ships, and the romantic age in which they played so large a part, will command the wistful admiration of men for all time.

James A. Farrell

INDIA HOUSE,
NEW YORK, 1927.

CONTENTS

CONTENTS

CONTENTS

PART III

CALIFORNIA CLIPPER SHIP ERA, 1850–1853

CONTENTS

CONTENTS

ILLUSTRATIONS

"Some like pictures of women and some likes 'orses best,
But I like pictures of ships, by Gum, and you can keep the rest,
And I don't care if it's North or South, the Trades or the China Sea,
Shortened down or everything set, close-hauled or running free;
But paint me a ship as is like a ship and that'll do for me."

FROM *Pictures*—BY CICELY FOX-SMITH.

COLORED PLATES *

*In the present edition, the original color plates are reproduced in black-and-white. In addition, the following appear in color on the covers: "Sovereign of the Seas," "Flying Cloud," "Lightning," "Westward Ho!" and "Great Republic."

xxiii

ILLUSTRATIONS

ILLUSTRATIONS

ILLUSTRATIONS

xxvi

ILLUSTRATIONS

PLANS, MODELS, MAPS, FACSIMILE LETTERS AND TELEGRAMS, ETC.

SOME FAMOUS SAILING SHIPS AND THEIR BUILDER, DONALD McKAY

PART I

AMERICAN TRADING AND EARLY PACKET PERIOD, 1810–1844

DONALD McKAY AT THE AGE OF 54 YEARS

From a steel engraving in the McKay Collection.

CHAPTER I

IN the early years of the nineteenth century, a wagon laden with potatoes came rattling down to the waterfront of the old Loyalist town of Shelburne, Nova Scotia. On a wharf it stopped with a jolt, and a lad, who had been sleeping on the piled-up load, awoke with a start. The little chap sat up and rubbed his eyes, then suddenly was all intent, for there, at the wharfhead, lay a Banks fishing schooner! The eyes that a moment before were closed in sleep now danced with delight, for they had been seeing ships in a dream. The lad on that load of potatoes was Donald McKay, and the world has since marvelled at the realization of his dreams!

Was his youthful imagination even then conceiving images of the *Flying Cloud*, the *Lightning* or that Queen of Ships, the *Great Republic?* Whether or not the lad dreamt any of these things, we cannot know. We do know that from his earliest youth Donald McKay was enamored of ships; and that this love grew and grew until his genius created and his skill constructed those superb specimens of marine architecture that won him world-wide renown.

Donald McKay, America's celebrated designer and builder of ships, was born at Shelburne, Nova Scotia, September 4, 1810. He was the eldest son of a farmer, Hugh McKay, whose father, Donald McKay, of Tain, Ross County, Scotland, was a British Army officer who emigrated to Nova Scotia after the close of the American Revolution. His mother, Ann McPherson, was the daughter of Lauchlan and Elizabeth McPherson, and she belonged to a family which had long been settled in Nova Scotia and had attained prominence and wealth in the professional and commercial circles of Halifax. She died in Boston, November 14, 1856, but both she and her husband, who survived her until December 30, 1871, lived long enough to see their son, Donald, attain fame and honors accorded few men of his time.

When about sixteen years of age, Donald McKay was compelled through necessity and an eager ambition to leave his Nova Scotia home to seek a career in New York. At that time, 1826, shipbuilding was the leading industry of the American metropolis and New York-built ships were second to none in the world. That city was the headquarters for the packet business between America and Europe and possessed the most and finest sailing ships on earth. It was natural, therefore, that our aspiring youth should gravitate toward that great seaport.

From a coaster that sailed from Halifax, after a long, rather stormy voyage, he landed in New York. It was not difficult for him to locate the shipbuilding area. The whole waterfront along the East River, from Corlaers Hook to about East Tenth Street, was one mass of busy shipyards, and he found ready employment, as a sort of day laborer in Isaac Webb's yard, located on the East River, from Fifth to Seventh Streets. This employment was only temporary, for he had decided to learn the ship-wright trade thoroughly, and to do this he must serve an apprenticeship. So he became "indentured" to Isaac Webb, appropriately called the "Father of Shipbuilders," because more successful master shipbuilders were grad-uated from the latter's yard than from any other place in America. Webb was a man of sterling qualities, ever considerate of the needs and feelings of his workmen, at a time when harsh treatment by bosses was the rule.

The life of an apprentice to a shipbuilder in New York, one hundred years ago, when this ambitious youth "served his time," would be unbearable to a modern mechanic. How fine that mill ground is evident from the apprenticeship indenture, which we now reproduce.

This Indenture Witnesseth, That Donald McKay, now aged sixteen years, five months and twenty days, and with the consent of Hugh McKay, his father, hath put himself, and by these presents doth voluntarily and of his own free will and accord put himself, apprentice to Isaac Webb, of the City of New York, ship-carpenter, to learn the art, trade and mystery of a ship-carpenter, and after the manner of an apprentice to serve from the day of the date hereof, for and during and until the full end and term of four years, six months and eleven days next ensuing; during all of which time the said apprentice his master faithfully shall serve, his secrets keep, his lawful commands everywhere readily obey: he shall do no damage to his said master, nor see it done by others without telling or giving notice thereof to his said master; he shall not waste his master's goods, nor lend them unlawfully to any: he shall not contract matrimony within the said term: at cards, dice, or any other unlawful game he shall not play, whereby his said master may have damage; with his own goods nor the goods of others without license from his said master he shall neither buy nor sell; he shall not absent himself day nor night from his master's service without his leave; nor haunt ale-houses, taverns, dance-houses or playhouses; but in all things behave himself as a faithful apprentice ought to do during the said term. And the said master shall use the utmost of his endeavors to teach, or cause to be taught or instructed, the said apprentice in the trade or mystery of a ship-carpenter, and the said master shall pay to the said apprentice the sum of two dollars and fifty cents weekly for each and every week he shall faithfully serve him during the said term. And shall also pay to him, the said apprentice, the sum of forty dollars per year, payable quarterly, for each and every of the said years, which is in lieu of meat, drink, washing, lodging, clothing, and other necessaries. And for the true performance of all and singular the covenants and agreements aforesaid, the said parties bind themselves each unto the other firmly by these Presents.

In Witness Thereof, the parties to these Presents have hereunto set their hands and seals the 24th day of March, in the year of our Lord one thousand eight hundred and twenty-seven.

> ISAAC WEBB
> DONALD McKAY
> HUGH McKAY.

Certain of the master shipbuilders erected, as a boarding house for their apprentices, a large brick building in Columbia Street, popularly known as the "Weary Wanderers' Hotel," and here our struggling apprentice and later on, his brother, Lauchlan, were domiciled for a time during their apprenticeship. From all accounts, it was the headquarters of considerable jocularity; and, we daresay, the future builder-owner of a fleet of the finest ships that ever sailed the main, often joined his companions, when they went out upon those festive nocturnal expeditions that made these precocious shipyard apprentices the terror of the neighborhood.

His apprenticeship fulfilled, the aspiring mechanic became, at twenty-one years of age, a full-fledged shipwright or carpenter; but it would be a mistake to suppose that this young man's labors were easy. He worked from sunrise to sunset; that is to say,—from half-past four o'clock A.M. in summer till half-past seven o'clock P.M., a period of fifteen hours,—for $1.25 per day. At eight o'clock in the morning he was allowed an hour for breakfast. At twelve o'clock he had two hours for dinner.

His supper came after the day's labor. The heaviest beams, which are now lifted to the stocks by steam or electricity, he carried on his shoulders, his bosses working with him, and usually not sparing their emphatic orders. Many hours would be consumed in the sawing of a piece of timber of live-oak that well-exemplified its name. Using a two-handed saw, one man would stand upon the beam, and the other below it in a saw-pit or ditch that had been dug to hold him, his face protected by a veil from the dust. Nowadays a circular steam, or electric-driven saw would go through such a beam in a fraction of the time it then took.

Eventually the wages aforementioned were raised, and a working day of ten hours ushered in. It was about this time that Donald McKay made a decided move. The "indenturing" had palled upon him. The so-called "slavocratic" conditions imposed by a shipyard apprenticeship, on a young man of his indomitable energy and natural mechanical talent, were too onerous to be long endured. He, therefore, decided to demand a release. To the credit of Isaac Webb, it must be said that he had somewhat anticipated such a move. Our young Nova Scotian had attracted his attention by his unusual industry and ability, and this competent master of men had already singled him out and entrusted him with responsible duties and far more important work than usually fell to the share of apprentices. Isaac Webb's

8

DONALD McKAY

At the age of 20, when serving as a shipwright apprentice in New York. From daguerreotype in the author's possession.

well-known honesty of purpose triumphed over the consciousness of his legal rights, and he allowed the apprentice-bound youth a free and clear release from the terms of his indenture. He forthwith entered the employ of Brown & Bell. Mr. Jacob Bell of this firm had not only formed a strong attachment for him personally but had realized his ability as a craftsman.

Thus it can be seen that the youthful Donald, although a stranger in New York and comparatively friendless, met with a fair measure of success early in his career.

Now his restless ambition received an impetus from a feeling that has carried many a man over otherwise unattainable heights. In other words, he fell in love—fittingly of course, with a shipbuilder's daughter.

At this time, 1832, Donald McKay was what may be termed a "free lance" shipwright of locally established reputation. Packet building was the best work of the New York yards at this time, and in the construction of the packets, especially the Liverpool, London and Havre packets, the majority of which were built in three yards,— Isaac Webb's, Brown & Bell's, and Smith & Dimon's,— the best shipbuilding talent of the day was employed. Regular work was to be had by young McKay, for his employers soon recognized his capability.

Democratic in their tastes and simple in their habits, New York's East River shipbuilding community had little social intercourse outside of their own neighborhood.

It was natural that Donald McKay should bestow his affections and marry within this immediate circle. Albenia Martha Boole was the eldest daughter of John Boole, who had been engaged in the shipbuilding industry for many years. Two of her brothers, also, were shipbuilders. She was an unusually talented, energetic young woman, possessing a good education. Brought up in a shipbuilding atmosphere, she not only understood much about ship construction, but could draught and quite expertly "lay off" plans for a vessel. This capable woman became the mentor and teacher, who imparted to Donald McKay not a little of his knowledge of marine architecture at the beginning of his career. Oftentimes his unceasing ambition caused him to feel the need of knowledge other than what he had acquired at Shelburne in his early youth,—a period so short as to cover only the fundamentals of an education. In brief, this man of extraordinary mechanical endowments, had possessed only what may be called a "primary education," previous to his courtship of Albenia Boole. But his was a will that triumphed over all impediments; he rapidly acquired not only an excellent knowledge of the rudiments of that education which had been denied him in his youth, but so well succeeded in the acquisition of learning that during his remarkable after-career, he consistently held his own with nearly all kinds and classes of people.

The young couple established themselves in a little

home of their own on East Broadway, considered then one of the finest residential streets upon the East Side of New York. Here their first child, Cornelius Whitworth, was born February 1, 1834. As the Boole family were well-to-do, the wife brought young McKay a tidy sum of money. She was economical as well as capable and industrious in her household, and those "hard times," too often encountered by young married people, did not mar their happiness. Then, again, his labors at the shipyards continued steady and wages were good.

Employed as a draughtsman by Smith & Dimon, whose shipyard lay at the foot of East Fourth Street, was a gray-eyed, dreamy-browed fellow, in his early thirties, who later exerted great influence over McKay's marine architectural productions. His name was John Willis Griffiths, and his genius revolutionized the science of merchant shipbuilding by the introduction of the first clipper ship model. He and the future designer and builder of the finest clipper ships that ever "slid down the ways"—the latter being at this time employed in an adjoining East River shipyard—naturally met each other; and then and there commenced a friendship which lasted for many years. Griffiths created no small sensation in New York shipbuilding circles when he attacked the predominating theory, that it did not matter how roughly a vessel entered the water so long as she left it smoothly behind her—the theory exemplified in the Baltimore

11

clipper's full round bows, practically flat forward floor and narrow stern. This daring innovator proposed a model of a knifelike, concave entrance, melting into an easy run to the midship section, where, instead of forward, he located the extreme breadth of beam. Thence this fullness of breadth melted again into the after end in lines almost as fine as those forward. In place of the codfish underbody, he gave his innovation a dead rise amidships. Later on, he carried this innovation into practice in designing the pioneer American clipper ship *Rainbow*. Donald McKay became, in time, his most famous disciple.

It should be borne in mind that during this period, packet building was the best work of the New York shipyards, as that city was the centre of the packet business between America and Europe. There were lines from other ports; but New York, the pioneer, always kept the lead, and had the most and finest packets. The frequency and regularity of their sailing were amazing. Towards the last there was a packet ship sailing every five days. No foreign vessels carried the mail in those days, as it was all given to American packets; and so great was the reputation of these vessels that they were regularly patronized, not only by Americans going abroad, but by West India merchants, Canadians, and even by the English officers of the large garrisons in Halifax and the provinces generally. For many years these swift sailing New York packets drove nearly all their foreign rivals out of the shipping business.

In the construction of these vessels the ablest mechanics were employed,—selected men, well-known as a rule for their capability. A large percentage of them were Americans, for there was then no place more patriotic than a New York shipyard. The rivalry among the various trans-Atlantic lines was keen, and it was this eager competition as much as anything else, that led to the continual improvement in the models, rig, workmanship and general excellence of American ships—each new vessel being expected to excel some rival or all the predecessors of its own fleet in some desirable quality. New York shipbuilders now found that more was required of them than at any previous period of their history. In order to hold their own and maintain the reputation of their yards, they were forced to study the scientific principles involved in the form of the hull and in the sparring of ships. They did not sufficiently know what made one ship bad and another good, and therefore began to study and experiment. They eagerly sought every source of information. Delicate tests were made with small models of different forms; the flow of waves away from the bow of a boat was investigated and every other conceivable point was carefully looked into. Even fishes were cut up and their shapes analyzed. Here in the centre of this vortex of study, experiment, and discussion, we find our young shipwright striving, with honest effort, to learn, not only fundamentals, but, also, the scientific principles that governed the noble art of shipbuilding.

Upon the recommendation of that eminent ship-builder, Jacob Bell, Donald McKay obtained employment in the Brooklyn Navy Yard, and there his exceptional ability led to his selection from among nearly a thousand men as foreman of a gang having important work in hand. Unfortunately at that time a strong native American party-feeling prevailed among the mechanics, and, because McKay was not born under the "Stars and Stripes," they bullied him out of the yard! However, within a comparatively short period of time, this "victimized alien" virtually invested those same "Stars and Stripes" with more maritime honor and glory than they have ever since received.

CHAPTER II

MR. BELL was still Donald McKay's friend, and now sent him to Wiscasset, Maine, to draft and superintend the building of some ships for New York shipping houses. This visit to the Eastern country opened up to his restless ambition a fertile field, —so far behind the New York shipwrights were the New England mechanics at that time in methods of construction.

The first place in Massachusetts, coming down the coast, where there were any shipyards in 1840, was Newburyport. Here, while visiting he finished the ship *Delia Walker*, of 427 tons, for John Currier, Jr. From time to time the owner of this vessel, Mr. Dennis Condry, came to watch the progress of the work, and he became greatly impressed with the unusual mechanical ability of the young foreman in charge of the job and also with the amount of work that he was able to get out of his men. Several years later, when opportunity arose, he was able to say a good word for that young shipwright and thereby brought about a turn in his career that eventually led to fame and commercial success.

On the completion of this vessel, John Currier, Jr., then considered Newburyport's foremost shipbuilder, was so pleased with the ability and energy of his young foreman, that he wished to bind him for five years of service. This apparently advantageous offer was refused, the refusal being due to the young man's eager desire to embark in business on his own account.

By strict economy Donald McKay had saved a little money, and when another Currier, William by name, offered him a partnership, he moved his family to Newburyport; and as a member of the firm of Currier & McKay, commenced his career as a shipbuilder.

The barque *Mary Broughton*, 323 tons, was built by the partners during 1841, followed in 1842, by the ship *Ashburton*, 449 tons, then the *Courier*, of 380 tons, which was really McKay's first production as a designer and builder of ships. She was built for Andrew Foster & Son, well-known merchants of New York, and commanded by Capt. W. Wolfe, who owned the then customary captain's share in the vessel. The *Courier* was built to sail in the "Rio" or coffee trade between New York and Rio de Janeiro, and shipping men in those days scouted the idea that such craft as were demanded by the competition in that trade could be built outside of New York and Baltimore. Her success, when sailing side by side with

"RIO" TRADER "COURIER," 380 TONS

The first ship built by Donald McKay in 1842, at Newburyport, Mass., for Messrs. Andrew Foster & Son, New York. From painting by Charles Robert Patterson, owned by Frederic de P. Foster, whose father and grandfather gave Donald McKay his first commission to construct a ship of his own design.

Joshua Bates, 620 Tons, 1844. Pioneer packet of Enoch Train's White Diamond Boston-Liverpool Line. Her superior construction so impressed Mr. Train that he urged Donald McKay to leave Newburyport and move to Boston. Picture from the MacPherson Collection.

the finest vessels afloat was little short of wonderful, and, as quick passages meant money then, her builder was at once brought prominently before the maritime public.

From "Some Recollections," an interesting biography of himself, by that scholarly sailorman, Capt. Charles Porter Low, is extracted the following account of a voyage he made in this little trader:

After being ashore four weeks I longed for the sea again and about the middle of January I shipped on board the *Courier*, Captain Wolfe, for Rio de Janeiro. The *Courier* was a small ship of about three hundred and fifty tons, very fast, and a beautiful sea boat, but after being on board the *Toronto* it seemed child's play to handle her royal and topgallant sails. Captain Wolfe was a very kind and pleasant man. He had good feed, and "watch and watch," with a very respectable crew of twelve men, four ordinary seamen and four boys, cook and steward and two mates, carpenter and sailmaker.

We left Sandy Hook with cold weather and a fresh westerly wind which, the second day out, increased to a heavy gale with a snowstorm. The ship under three close-reeled topsails and reefed foresail ran before it, and she did beautifully, and we soon ran across the Gulf Stream and had good weather.

We had a very short passage of thirty-eight days to Rio de Janeiro, and almost as soon as we anchored and as soon as the ship was entered in the Custom House, we went to work discharging our cargo of some nine thousand barrels of flour.

Later on he writes:

We had a longer passage home, of some forty-eight days. Light winds and calms and a heavily laden ship delayed us. The voyage was one of the pleasantest I ever made.

The firm of Currier & McKay was soon dissolved, models and moulds being divided (with a saw) in two equal parts. McKay then connected himself with William Pickett, and, under the firm name of McKay & Pickett, constructed the New York packet ship *St. George*. This vessel, of 845 tons burden, was the pioneer ship of the Red Cross Line, or the St. George's Cross Line, as it was also called and advertised. Curiously enough, the Line is at the present time chiefly remembered through a clipper ship built ten years later, the famous *Dreadnaught*, which commanded by doughty Captain "Sam" Samuels richly earned her sobriquet of the "Wild Boat of the Atlantic."

Newburyport in the early forties certainly was a busy shipbuilding town. Its premier element socially as well as commercially, was a refined, rather reserved community, consisting of a few large families, all of which were engaged in shipbuilding or kindred trades.

The charming qualities of his talented wife and his own interesting and attractive personality, together with his unusual mechanical knowledge and ability—the last much appreciated in a town where everybody talked "ship,"—made the young couple welcome in the best social circles of Newburyport.

It should be noted here that Donald McKay had a very high ideal of womanhood and was always drawn to cultivated and refined people. He appreciated education all the more because he had received little schooling himself. He possessed a fine ear for music, and, in his younger days, used to play the violin for dancing reunions and other social occasions. In later years he became too much engrossed in his immense shipbuilding operations to do this.

Among the prominent and most respected men of Newburyport was Orlando B. Merrill, a noted shipbuilder who built the sloop-of-war *Wasp* which captured the *Frolic* after a terrible combat in the war of 1812. He it was who first conceived, in 1794, the water-line model which practically revolutionized the science of shipbuilding. Previously ships had been built from skeleton models composed of pieces that showed the frames, keel, stem, and stern post, but were of little use in giving an accurate idea of the form of the vessel, while much time and labor were required to transfer the lines of the model to the mould loft. Merrill's water-line model, composed of lifts joined together, originally by dowels and later by screws, could be taken apart, and the sheer, body, and half-breadth plans easily transferred to paper, from which the working plans were laid down in the mould loft. This mechanical genius maintained a long and intimate friendship with Donald McKay; and the fact that McKay's models were and are still regarded as masterpieces of the art and science of shipbuilding, may be attributed to the instruction he received from the inventor himself.

In 1844, the packet ship *John R. Skiddy*, 930 tons, was built by McKay & Pickett, for account of Captains William and Francis Skiddy of New York. She was commanded by Capt. William Skiddy, one of the most famous and popular packet masters engaged in the New York-Liverpool trade. The *Skiddy* soon showed that she was the work of an able mechanic, warranting these same two nautical gentlemen to afterward employ her builder in the construction of other ships at his yard in East Boston.

When finishing the ship *Delia Walker* in the yard of John Currier, Jr., McKay had attracted the attention of her owner, Dennis Condry. This gentleman and true friend, when traveling abroad afterwards, mentioned the young man's acquirements to Enoch Train, at that time the leading shipping merchant of Boston. He extracted, at the same time, a promise from Mr. Train that before contracting for his contemplated line of packet ships, which were to sail in the European trade, and which later became famous as the White Diamond Line of Boston-Liverpool packets, he would, at least, visit this rising young shipbuilder. To fulfill that promise, Mr. Train visited Newburyport, and the contract to build the ship *Joshua Bates* of 620 tons, was made in an hour. It was "flint and steel" when these two master-minds, merchant and mechanic—struck together.

The *Joshua Bates* completed—a masterpiece of shipbuilding—Newburyport became too small for its creator's inordinate desire for attainment. Upon the day when this splendid vessel was launched and floated safely upon the Merrimac River, Enoch Train grasped Donald McKay by the hand and said to him, *"Come to Boston; I want you!"* This invitation was followed by satisfactory arrangements for financing shipbuilding operations, in East Boston, where McKay decided to locate his shipyard. Dissolving his pleasant and profitable partnership with Mr. Pickett, the young shipbuilder, now his own master—in realization of his early day dreams,—soon left Newburyport for Boston.

PART II

AMERICAN PACKET SHIP PERIOD, 1845–1853

CHAPTER III

THUS it was that Donald McKay, at the age of 34 years, realized, in no small measure, the result of his capabilities. His great ambition, to establish his own shipbuilding plant where he could construct craft of his own design, was not now a vain or visionary conception! He lost little or no time in commencing the work of preparing a shipyard at the foot of Border Street, East Boston.

He took his family to a house in Princeton Street but in about a year's time he erected the residence on White Street, which remained his home throughout his career. Here he entertained some of the most famous men of his time, and it was customary after each launching to repair there and partake of his hospitality. Designed, constructed and finished with the same skill and attention to mechanical detail that he always exercised in the building of his ships, this house remains still (in the year 1928)

a memento upon terra firma of his prowess, whereas of
the many beautiful specimens of marine architecture his
genius created none remain.

The end of the War of 1812, the dawn of 1815, found
American Shipping and its corollary industries moribund
—every port, from the Penobscot to the Chesapeake,
choked with idle, dismantled or rotting fleets. Merchant
shipbuilding in the United States, as well as abroad was
no more than a rule-o'-thumb trade. Naval construction
occupied only a slightly higher plane.

A special class of ships which now grew up were the
sailing packets, or vessels carrying both passengers and
freight. Those which cleared from port on regular days
in each month and ran back and forth between special
points only, were called "liners." These "liners" were
conspicuous for sailing at their advertised time, wind or
no wind, gale or calm. When carrying the mails they
were allowed nine to ten weeks for the round passage on
the trans-Atlantic route. As commerce increased between Europe and America these "liners" were enlarged
and greatly improved both in build and in their accommodation for cabin and steerage passengers. In this
respect, and especially for the latter class, they excelled
the transient vessels.

As compared with the comforts emigrants to America
have now in crossing the Atlantic, they had a hard time
of it in those days of sailing ships. The earlier steerage

passengers had to provide themselves with food for the voyage. They had to cook their own victuals, weather permitting, at an open galley fire on the waist-deck, and many lively scenes were witnessed around the galley-fire as each emigrant struggled to get his or her pot upon it! Often, through a miscalculation as to its duration, they suffered a great deal from their supplies falling short. In order to prevent this, an act of Parliament was passed, obliging all vessels carrying emigrants to supply them with a daily ration.

Amusing indeed were the scenes on the stated days on which rations were served out to these adventurers and hardy pioneers of the West, as one of the mates weighed out to them their allowances of flour, beans, rice, etc., and the ship's carpenter doled out each person's allowance of water.

Sleeping accommodation was of the rudest kind, put up by the carpenter and his mate, according to the number of passengers, a few days before sailing. The whole arrangement and government of the place fell upon the ship's carpenter, who also had to see that no one exceeded his or her proper allowance of water and fuel, also that the ship was not set on fire by men smoking below, etc.

In stormy weather their sufferings may be imagined when the hatches were battened down, and the scores and sometimes hundreds of men, women and children were often kept for days in a stifling atmosphere. When the voyages were prolonged by adverse winds, many of these sailing ships became floating pest-houses. Ship or typhus fever, small-pox or some other disease, fed by the surroundings of their victims, slew scores and even hundreds of them during the voyage, or when detained in quarantine on arrival. The earlier packets carried no surgeon or doctor; and unless one turned up by chance among the passengers, this duty in the steerage fell upon the carpenter, who dispensed all medicines required, which were served out to him, according to the mate's advice. An old, but very large "Dictionary of Domestic Medicine" was consulted in doubtful cases. Generally speaking the carpenter ("Chips") was also the ship's dentist.

Still in the face of all discomforts, dangers and drawbacks, emigrants went in thousands across the ocean to seek homes in the western land of promise, and by their Anglo-Saxon enterprise largely contributed to the growth of our great American Republic and the Dominion of Canada. Thus was built up that over-sea traffic, under which Atlantic navigation prospered and grew.

Fixed dates of departure were generally announced with the addition of the well-known proviso "Wind and weather permitting." This only applied to transient vessels; the "liners" sailed as advertised, fair weather or foul. A large number of trans-Atlantic lines came into existence as a natural outgrowth of the rush of emigration from Europe to America and the general expansion of ocean travel and trade. The carrying of passengers was a profitable business, and there was considerable competition among shipping merchants to get the largest share. None but the best and finest vessels could be used in this business, and the old-fashioned freighting ships, with their small cabins and houses, underwent a considerable change to adapt them to the new state of affairs. A great many shipping houses in Portland, Boston, New York, Philadelphia, Baltimore, Norfolk, Charleston, and New Orleans put their money into ships especially built for the passenger service, and ran them in regular lines to all the ports abroad and on our own coast whither trade and travel chiefly tended. In the coasting trade brigs, schooners, and barks were used, but in the packet service to foreign ports barks and ships only were thought of. The latter were vessels of the largest size, handsomely built and sumptuously fitted out, and carried from 600 to 1,000 persons and 1,000 tons or more of freight. The lines to Liverpool, London and Havre, often comprised 12 or 15 vessels, each one of the finest specimens of marine architecture afloat. When the clipper era began, sailing packets improved in speed and made trips across the ocean to Europe and to Australia, which took the steamships years to surpass.

When ships of five, or even three hundred tons, were not thought too small for the work, it was something worth talking and boasting about to make an Atlantic voyage. Before venturing upon it, men weighed its probabilities, and solemnly made their wills in view of its uncertainties. Nearly three months may elapse before their friends or expectant legatees could be informed of their safety or fate. While there were fixed days of sailing for the packets, there were not, as now in the case of our Atlantic steamers, fixed days of arrival. In fact, the sailing packet might reach Halifax (Boston or New York) in three weeks, or it might be in seven, for everything depended upon the weather and winds they encountered in making the voyage.

The qualities desired in a packet ship were strength, speed, stability at sea, ease of handling, easy rolling, beauty of model, and comfort in the passenger accommodations. It must not be supposed that these were all attained at one bound; on the contrary, the best good general model for the packet ship, and the best sizes and dimensions of timbers, were reached only by patient study and slow degrees. A great many bad ships were built before all the questions that interested the building world were decided; but it was as a final result of many years of study and investigation following the war of 1812, that the sailing ship reached substantial perfection as an ocean carrier.

Down to 1849 packets were either one or two-decked vessels, with a poop-deck aft and a top-gallant forecastle forward. Those in the service to Europe were of from

900 to 1,000 tons register. The cargo was stored in the lower hold, some of the light freight going between-decks if the cargo was a large one. The between-deck space aft was divided into cabins for the passengers; the middle portion was fitted up with kitchens, pantries, etc. The steerage passengers and crew were forward. On deck were houses for the crew and various officers, with others which served as vestibules to the apartments below.

The cabins of these "liners" were so largely in advance of those of any other ships of the day, that they were looked upon as superb. Eighty cabin passengers were considered a good list.

The steward's department fully met all the requirements of cabin passengers, who did not expect first-class hotel accommodation at sea, as they do now, but were contented with a fare which would not today be tolerated by trans-Atlantic voyagers even in fourth-class steamers. Biscuits were then used as the sea substitute for bakers' bread, those for cabin passengers being finer in quality than the "hard tack" supplied to the crew. In course of time baking hot rolls for the cabin was introduced, and regarded as a wonderful improvement in Atlantic catering.

The freight consisted chiefly of Virgin Turpentine and Pitch for ballast; and various food stuffs, tobacco, lard, cheese, oil-cake, woods and staves. Such a cargo was paid for by weight and measurement—barrels by the piece, wheat by bushel, cheese by the ton, and tobacco by the hogshead. From $5,000.00 to $10,000.00 freight money and from $2,000.00 to $5,000.00 passage money were the usual returns of an outward voyage, although occasionally these sums were larger. All freights were insured for full value and usually the vessel was also thus insured. From $20,000.00 to $30,000.00 insurance would be carried on the hull of each packet, the yearly rate being from eight to twelve per cent, paid by giving a note due in twelve months. For the cargo, according to the season of the year, from two and one-half to four per cent.

In their day packets were as well known for their respective size or speed as our crack steamships are now. They also had their celebrated and popular captains, of whom the passengers who sailed with them had always something to say, and were expected to say it. The rapid passages excited great interest, especially when one of the ships broke some outstanding sailing record. It was acknowledged on both sides of the Atlantic that the American packets carried the best officers and crews.

The comings and goings of these wind-propelled vessels were ordered exactly—fair weather or foul, they sailed as scheduled. The skill of their driving commanders reduced a passage of terror, of from a month's to three months' duration, and frequently longer, to one of comparative luxury and a definite number of days— from fourteen to twenty. Beyond doubt or question during the period from, say, 1820 until about 1848, American packets seized to themselves a monopoly of the North Atlantic trade. The Black Ball, Red Star, Swallow Tail and Dramatic Lines of New York and Enoch Train's White Diamond Line of Boston, and Cope's Line of Philadelphia were the boast and pride of the nation.

While there has always been an effort, even in ancient times, to improve the models and speed of vessels, still the adoption of steam, in place of sail, for trans-Atlantic communication was delayed fourteen years. This was notwithstanding the successful passage of the pioneer steam vessel, *Savannah* which created a popular craving for the application of steam to ocean navigation. It was not until 1832 that the Atlantic was crossed by a regularly built steamship. While the splendid service of the American sailing packets was a deterrent factor, it is interesting to note the strong prejudices that largely contributed against the application of steam as a motive power for these ocean voyages.

In the many public discussions, Dr. Lardner and other scientists took the negative side and demonstrated, chiefly to their own satisfaction as theorists, that ocean steam navigation was impracticable. The elaborate objections which were brought up to support Dr. Lardner's views, looked at in the light of present triumphs of ocean steam navigation, are most amusing. But as they are referred to now as showing the prejudices against it, we give our readers the following concise epitome of them.

His first and leading argument against the practicability of Atlantic steam navigation was based upon the physical condition of that ocean, such as "the atmospheric currents called trade winds, which, as they approach the equator, produce calms, interrupted by hurricanes, whirlwinds, and other violent convulsions." The difficulties of the Gulf Stream, and the weather of the zone of the ocean marked out by it, was, he declared, extremely unfavorable to steamship navigation.

Then, in his view, the long swells of the Atlantic, caused by the prevailing westerly winds, were more disadvantageous to a steamer than the short chopping waves of inland seas. He described the former as masses of water "hurled with accelerating momentum over a tumultuous confluence of water 3,000 miles in compass," which an immense vessel, forcibly impelled by opposing steam-power, could neither successfully elude nor safely encounter." He then, in connection, we presume, with the furnaces of the boilers, pointed to the "calamity of fire, and to the danger of collision with icebergs, in the latitudes which steamers must necessarily pass through in crossing to Halifax and Boston." To these difficulties and dangers he added the coating of the boilers with soot, which would impair the conducting power of the iron of which they were made, as well as the liability of leakage through the continued motion of the machinery during the voyages. The anxiety and fatigue of engineers and firemen leading to their neglect of duty was amongst these "a priori" arguments used by this learned philosopher against the possibility and probability of trans-Atlantic steam navigation.

It must be conceded that the valiant little American packets, though their day was not a long one, played an important part in the making of this nation, for in conveying their multitude of nationalities across the Atlantic these ships helped populate the United States of America; and they gave the down-trodden peoples of Europe a chance to become useful citizens in a new world. Their wonderful development, and the events that transpired during a progressive period, when ships of only 300 tons were supplanted by "fast liners" of over 2,000 tons, form an interesting study. The part that Donald McKay took in it all, as packet after packet was launched from his yard, each enlarged and greatly improved, is now told exhaustively for the first time.

CHAPTER IV

"WASHINGTON IRVING," 751 TONS
LAUNCHED SEPTEMBER 15TH, 1845

W HEN Donald McKay, in 1845, laid the keel of the Boston-Liverpool packet ship *Washington Irving* for Train & Company's White Diamond Line, he gave birth, at his new shipyard in East Boston, to the first of a fleet which contained many shipbuilding gems, all of which bore the unmistakable impress of his artistic taste in design and thorough honesty in carpentry. "Excelsior!" was his motto; the Best, and nothing but the Best for each and every craft for which he stood sponsor.

One cannot help admiring the daring hardihood that impelled Enoch Train to start his celebrated line of sailing packets and to commission Donald McKay to build them expressly for the trans-Atlantic passenger, freight and mail service. He had to contend not only with the keen rivalry of the New York Packet lines, for the Lords of the Admiralty, who had charge of the Royal mails, sometime previously had contracted for the conveyance of these with Mr. Samuel Cunard, who then brought into existence

SHIP "WASHINGTON IRVING," 751 TONS

First packet built at East Boston in 1845, for Train's Boston-Liverpool Line. From a drawing by Donald McKay at the Peabody Museum, Salem, Mass. Memorandum at bottom is in the shipbuilder's own handwriting.

PACKET SHIP DECK FARM

Here one can see the ship's long-boat, surrounded amidships by the cowhouse, caboose for cabin-cook, sheep, pig and poultry pens, with spare yards, topmasts, etc.

From illustration by Robert C. Leslie in the McKay Collection.

the British and North American Royal Mail Steam Packet Company. Of course American packet ships could still depend upon carrying the mails from this country. Up to 1846, when a regular international mail system was organized, it cost about a dollar to send a letter from America to Europe; consequently considerable revenue was derived from mail-carrying contracts.

Furthermore, owing to the dull condition of trade at Boston when she was first ready for sea, the *Joshua Bates*, pioneer ship of the Train line as organized about a year previously, had been sent to Mobile to procure a freight. Her performance on this voyage however was highly satisfactory, and Liverpool traders began to appreciate Enoch Train's efforts to merit their co-operation in establishing his line.

Among other improvements that were made in Train's packets was the elimination of that picturesque, though loathsome and noisy, feature heretofore found on the deck of even a line-of-packet-ship—the Old Ship Farm.

Always securely stowed amidships, well lashed down and housed over, the ship's clumsy long-boat was full of live provender, and as it lay on the deck, looked more like a working model of Noah's Ark than anything likely to save life at sea. Upon the first floor or main deck quacked ducks and geese; while above them, literally in the cockloft, were coops for chickens, etc.

The long-boat was surrounded by various small farm

buildings required to house the live stock for supplying the cabin table, the most important being the cow-house, where, after a short run ashore on the marshes at the end of each voyage, a well-seasoned animal of the snug-made Alderney breed, chewed the cud in sweet content. Preserved milk was unknown in those times; and the officers of a passenger ship would rather have gone to sea without a doctor, to say nothing of a parson, than without a cow and some nanny-goats. The ship's cow and her health was always a most important matter, and it is related that upon one occasion, after a long spell of very bad weather, one of these creatures fell off in her supply of milk and was brought around again by a liberal supply of nourishing stout, wisely prescribed for her by the ship's doctor.

Pigs—or as the old seamen usually called them, "hogs" —always proved a thriving stock on a ship farm. Next to the pig, goats were the most useful stock. These animals soon made themselves at home on shipboard; they had good sea-legs, and were blessed with an appetite that nothing in the way of tough fibre was too much for, from an armful of shavings to an old newspaper or log-book.

It was not, however, always practicable to turn in sheep to feed with pigs at sea, for the last-named animals were apt to develop a taste for a good live leg of mutton after a few weeks afloat.

Truly in those Board-of-Tradeless days, a ship was more like a small bit of the world afloat than it is now.

One can imagine the noisy confusion that must have reigned aboard one of these little packets on sailing day. Ducks, geese and poultry in general always sympathized with excitement near them, while pigs and even sheep, thrown together for the first time, had a noisy way of their own. At intervals, even the old cow bemoaned her lot in life. But all this clamor was soon lost in the rapid click, click, click of a capstan, and regular tramp round and chant of the red-shirted sailors. The chorus of "Good Morning, Ladies All," now swelled quaintly up at intervals above the other sounds, as the crew reeled in the warp by which slowly, but very surely, the ship was hauling out of dock.

The *Washington Irving* was strongly built of the best materials and beautifully modelled, her lines truly rounded, notwithstanding the sharpness of her ends; in fact, her lines were smooth and regular as those of a pilot boat. She was one hundred and fifty feet, ten inches in length, thirty-three feet wide, and twenty-one feet deep. The arrangements of her decks followed along lines that were later adopted in the construction of other Train packets, viz., she had a full poop, a top-gallant forecastle, a large house amidships before the main hatchway, and a deep waist.

Although registering only seven hundred and fifty-one tons, this packet was quite heavily sparred, especially for the period of her construction. American shipbuilders had not yet conceived the better plans for sparring vessels that they adopted later, more particularly during the Clipper Ship era. The vessels then employed in the trade between Boston and England were of three to five hundred tons, carrying cargo only as dead weight just sufficient to enable them to carry their canvas in good sailing trim. Donald McKay built with a view to the combined carriage of cargo and passengers, which enabled Train & Company to meet an increasing passenger and carrying trade.

Under command of Captain Eben Caldwell, the *Washington Irving* was promptly placed on the berth for Liverpool.

In *Young America Abroad*, under the title "Youthful Speculations," George Francis Train, who at one time served as a freight clerk with Enoch Train & Company, tells a more or less moralizing story:

I can well remember the fatal result of my first shipment of onions to Great Britain. I saw everybody shipping them, and believed a fortune was to be made, and I worked days, thought at my meals, and dreamt nights, until the bills of lading were signed at 2s. 6d. freight for twenty-five barrels silver skins, marked "T" (in a diamond) to Liverpool, on board the good ship *Washington Irving*. They were picked by hand and packed in the cleanest of barrels, and coopered with scrupulous exactness, and paid for, from my clerkship salary. (I went without a new suit for Sundays all that summer.) 'Twas awful suspense; my letter of instructions was carefully written, not a word too many, simply advising the quality, and using the words "prompt sales" and "prompt remittances." Four months went past and I was about ordering Menard to prepare the aforesaid suit, when lo! a letter came. 'Twas sealed, and bore that well-remembered stamp, B. B. & Co. (Baring Brothers & Company). I broke the seal; I saw the words "dull market," "regret," "perishable article"—and debit of £3 17s. 9d! That was my first and last shipment of onions to Liverpool! I stopped the order for the coat, and practiced economy until Richard was all right once more.

Until about 1852, when she was sold in England, the *Washington Irving* remained on the Boston-Liverpool packet route, doing valiant service in maintaining that so-called "Atlantic Shuttle"; and many prominent American and Canadian families of wealth and social standing today, can, if they would, trace the entrance of an ancestor into the land of promise, by the grace of this as well as other Train and Company packets.

CHAPTER V

"WE need an exceptionally attractive-looking ship to secure the passenger service in Boston and at Liverpool,—a craft not only A-1 in construction but one which will attract cabin passengers on both sides of the Atlantic and divert the European emigrant trade to our ships," said Enoch Train, as Donald McKay entered his counting-room at 39 Long Wharf, one morning in the spring of 1846. Then turning to Captain Alden Gifford, who superintended the building of his ships, Boston's leading merchant and ship-owner continued: "And this packet must be finished, fitted and decorated in the most modern manner." "Mr. McKay," he went on, "put plenty of ornamentation upon her fore and aft; build her about one hundred tons larger than the *Washington Irving*. She must be fast, very fast, and able to outsail any of the New York packets."

Thus Donald McKay was commissioned to build another splendid vessel which served as a pioneer in the

42

development of Train's trans-Atlantic passenger service. This staunch packet ship was profusely decorated, her figurehead ornamenting the bow being a veritable creation of an art almost unknown today: trailboards, the hanging knees under the catheads and the end of each cathead, etc., were ornamented with appropriate carvings, beautifully painted and gilded. Astern, too, she was extensively decorated, carved figures, indicative of her name, together with the American and British shields, flags, tree branches and other devices, being displayed in one gorgeous mass.

The *Anglo-Saxon* on the keel was 147 feet long, on deck 158, and over all 162 feet 3 inches; breadth of beam 35 feet, 3 inches; depth of hold 21 feet.

We will now glance at her on deck, and commence forward. Upon the topgallant forecastle, there was a splendid capstan of locust with a mahogany drum head, brass rim and composition pawls, and there was also another of similar materials upon the poop. Half way between the main rails and the water ways, there were rackrails, extending from forward aft to the gangways. Upon the house amidships, the long boat was stowed bottom up. She also carried three other boats, one from the stern, and one from each quarter.

She possessed unusually good accommodations for steerage passengers. Her between-decks were admirably adapted. They were lofty, well-ventilated, and the communication with the upper deck could always be open, as the hatchways forward and aft had ample protection from the sea and rain, the first being under the house amidships, and the other abaft the mainmast, under the poop. In each side there were four circular air ports with plate glass lights in them, each 8½ inches in diameter, set in composition frames, and secured with screws, so as to be air-tight when in. Forty-five feet of the forward part of the vessel were temporarily fitted for the accommodation of steerage passengers, and contained 96 berths, which were designed to be permanent.

As so large a proportion of the cost of this packet covered ornamentation, we furnish some details which convey an idea of the extravagance displayed to make her look attractive.

A full figure of an Anglo-Saxon monarch of the Middle Ages crowned and equipped for the field, ornamented her bow. His left foot was advanced, and his left hand rested on the upper edge of an oval shield, placed perpendicularly alongside of him. His right hand rested in his girdle. By his side he wore a sword, and his body dress was similar in fashion to that of the Scottish Gael. This figure was painted white, relieved by appropriate gilding.

The gangway boards were ornamented with lions, eagles, armorial bearings, etc., carved upon them, and the foremost stanchions of the poop bulwarks were also set off with carved standing knees.

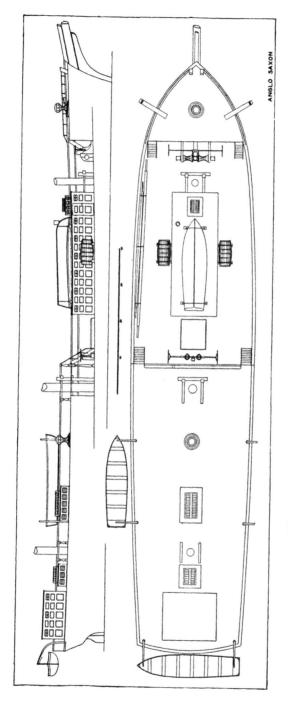

ANGLO SAXON

DECK PLAN OF PACKET "ANGLO SAXON"—894 TONS—1846

Drawing made by Alfred S. Brownell from description of this vessel in *Boston Post*, September 2nd, 1846.

Over the first tier of stern windows there was an extensive arch of carved work, in the centre of which were displayed American and British shields, side by side, guarded by an eagle and a lion, and surmounted by an unfurled flag and a liberty pole and cap. On the starboard side was an Irish warrior on horseback, bearing aloft in his right hand a spear; and on the opposite side an English chief, also mounted, and cased in mail, brandishing his sword. Below these, on either side, towards the archboard, there were branches and other devices, the whole being tastefully decorated with gilding, and painted white. The pilasters of the first tier of stern windows, and the moulding above the archboard were painted white, and her port of hail, "Anglo-Saxon, Boston," was carved into the archboard, and painted white, on a black ground. Her name, in gilded letters, carved in black ground, was also on each quarter of the poop bulwarks.

As was then customarily done with American-built ships, the *Anglo-Saxon* was to be either coppered or sheathed shortly after arrival at Liverpool.

An innovation, which has successfully stood the test of years—salt pickling—was here adopted by Mr. McKay. For the purpose of preserving the keelsons, before they were placed, circular grooves 3 inches wide and 2 deep, were cut into them the whole length, that is, the upper part of one keelson contained a semi-circular groove, which was matched by another of the same size in the base of the upper keelson, thus making the two, when joined, a circular tunnel. As there were three keelsons, of course there were two tunnels. These were filled with salt pickle, and had connecting pipes in the hold, through which more could be poured whenever the timber absorbed that already in. This plan, we believe, originated with Capt. R. B. Forbes, and was first applied by him in the construction of the ship *Massachusetts*.

It is but justice to add that the parties most interested in the *Anglo-Saxon*, the owners, were so well satisfied with her that they contracted with Mr. McKay to build another packet ship of larger dimensions.

The end of the *Anglo-Saxon* was tragically swift. Returning to Boston from Liverpool, upon her second voyage, under command of Captain Joseph R. Gordon, she was lost off Cape Sable late in the winter of 1846. Most of her passengers and crew were saved, after suffering severe hardships. Losing this costly vessel would have deterred most men, but Enoch Train kept right on with his plans to maintain a packet service between Boston and Liverpool that existed and prospered for many years afterwards.

CHAPTER VI

"NEW WORLD," 1404 TONS,
LAUNCHED SEPTEMBER 9TH, 1846

THE *New World*, when completed at McKay's yard, was another wonder. American packets were slowly increasing in tonnage, and this was the largest sailing vessel afloat. Furthermore, she was the first three-decked merchant ship ever built in the United States, and a notable accession to the New York packet fleet, which included the *Roscoe, Independence, George Washington, Pennsylvania, Patrick Henry, Ashburton*, and the well-known *Henry Clay* popularized in a chantey which was sung on both sides of the Atlantic.

At her launch, an event that had been looked forward to for some weeks, visitors were in town from the back country and from along the coast to see such a magnificent and hitherto unparalleled specimen of American marine architecture bid adieu to "terra firma." From New York, the night before, a delegation had arrived, which included her owners.

Launchings then, while not regarded as the social function they may be nowadays, were witnessed and appreciated by persons possessing keen interest and knowledge of the craft consigned to its watery element. Until a later period, it was not customary for women to name ships.

It was a day of anxiety to the builder until the ship had been successfully launched. He had so much at stake: the ways might be insufficiently greased; the chains beneath the vessel might break; she might tumble over on her side; she might acquire momentum enough to drive her into the opposite bank of the river, or collide with some nearby craft—oftentimes excursion boats filled with picnickers who came to witness the launch. If the weather was bitterly cold, he was fearful of the tallow freezing on the ways, so that when he had given the word to knock away the dog shores and the vessel moved (sometimes so slowly to his tormented mind!) down the smoking ways and plunged into the dark, forbidden-looking, icy waters, his joy and relief knew no bounds. If his workmen, who had labored hard upon the craft for months, now became more or less under the influence of liquor; who, in all fairness, could say them "Nay"?

The proprietors of traders, packet and clipper ships always insisted upon giving the workmen a "blow-out," and usually paid the bills for the biscuits, cheese and rum punch, etc., and also for the champagne drunk by the guests in the mould loft. It is said that Donald McKay did not like the saturnalia which these occasions often invoked, but that, be it remembered, was due to his interest in the welfare of his workmen and their families. The day after a launching many of his mechanics would be unfit for work or remain away from the yard altogether.

From a simple and impressive ceremony, usually performed by the foreman of the shipyard or an old faithful employee, launches grew in importance, as ships increased in size and monetary value. The fame of a shipbuilder's productions exercised no little influence, so that more dignity and honor, naturally, befell the "christener"—thus women and prominent men became ship sponsors. Launches came to be regarded as social functions, pavilions being erected to accommodate the many guests invited by the builders as well as the owners.

A slight jar, a rush to the sides, the roar of a cannon or an old-fashioned musket, loud huzzahs from outsiders, and the well-regulated cry of "Thar! Thar! she goes"— all these took place as the last block was cut away, and the vessel glided rapidly along the ways into the sea! Bowing obeisance to the spectators on the shore, she gracefully travelled about twice her length and was stopped by the properly heaved anchor or otherwise!

The crowd began to separate immediately after the flag-bedecked ship is off, and the rush for the various entrance-gates of the yard was tremendous.

The *New World's* first master, Captain William Skiddy, and his brother, Francis Skiddy, of New York, who some two years previously gave Donald McKay an order to construct the ship *John R. Skiddy* at Newburyport, had contracted with him to build this splendid vessel for the Liverpool trade. Previous to her first voyage, however, they sold their entire interest to Messrs. Grinnell, Minturn & Co., owners of the Swallow Tail Line, which ran packets from New York to Liverpool, under a blue before white swallowtail flag. They used for their London Line a red before white swallowtail. This well-known shipping firm probably had more ships built for their account than any in America, and the success achieved by the *New World*, which had been purchased by Moses H. Grinnell personally, proved advantageous to her builder in after years.

American packets were beyond dispute the finest and fleetest vessels afloat during the forties. New Yorkers had the exclusive and distinguished benefit of a fast weekly sailing service to Liverpool, semi-monthly packets to London and excellent service to Havre. Nowhere in the world were such splendid ocean-going vessels obtainable.

Cabin passage from New York to London and Liverpool cost $140; and to New York from London and Liverpool 35 guineas; a cabin passage to New York from Havre $140 and from New York to Havre the same. This included provisions, wines, beds, etc., so that passengers had no occasion to provide anything.

In *Reminiscenses of the Merchant Marine*, Captain

W. W. Urquhart gives the following interesting account of a New York packet departing from her pier along the East River waterfront:

When leaving port these ships were dressed with every flag in the code. The pier head was lined with friends bidding their last adieu, as with the steamers of today. The tug whistled, the moorings were cast off, the gun was fired, and the good ship was headed for her port of destination in charge of the pilot. The next thing to be done, when the ship was fairly under way, was to choose the watches and get the crew in order. All hands were called aft to the main capstan and lined up against the rail. The second officer, commanding the Captain's watch so called, has the first choice. In the meantime a trusted man of the after guard rummages the fo'castle for liquor—and I will say right here, whiskey, from fo'castle to the cabin, was responsible for many of the tragic scenes of these packet ships, especially on leaving port. The watches chosen, an equal number in each watch, it was my custom to make a short address, telling the men what we expected of them and what they might expect from us, etc. The usual speech, however, was: "Every man Jack of you relieve the watch at the capstan in five minutes. Be ready for a call when wanted or you'll catch hell. Now masthead the topsail yards and be damned to you!"

"Aye, aye, sir."

"Starboard watch aft—port watch forward!"

The Irish famine occurring about this time, the freight upon a barrel of flour rose to five shillings and the *New World* at once commenced making money for her owners. Upon her arrival in Liverpool, this packet was much admired, and, before leaving that port, received the honor of a visit from the Prince Consort. Later, in London and other seaports, she established the reputation of Donald McKay as a shipbuilder equal to the best.

Under command of Captain William Skiddy, who had previously commanded the Havre packet *Henry IV*, the *New World* soon became one of the most-deservedly popular American packets, which then rendered illustrious service to the growth of New York City as the commercial metropolis of the Union. She went out from New York to Liverpool in seventeen days, although becalmed about two days,—one of the fastest trans-Atlantic passages, considering the bad season of the year, ever made by a vessel of the same class.

No exertion was spared to avail of every breadth of fair wind, and of every "slant." Sail on these Swallow Tail "liners" was carried on so long as compatible with safety to ship or spars.

One of the most valuable qualities of a packet ship commander was weather wisdom and ability to correctly forecast the strength of an impending gale.

This little incident (a very common one, by the way) occurred in the "Roaring Forties" aboard one of these Liverpool packets bound to New York.

> I tell the tale as 'twas told to me
> By a tattered and battered son of the sea.

The Captain had remained on deck, watching his canvas since before noon, his own judgment as well as a falling glass admonishing him that sail must be shortened before very long; and at last, reluctantly, he ordered all hands called to double-reef the main topsail; there was a heavy sea on, into which the ship continually pitched knight-heads under, and the cross-jack and the spanker were then taken in to ease her. The gale rapidly increased in fury, accompanied by heavy squalls of rain and sleet, with a dangerous cross-sea. At eight bells all hands were called, and the watch, instead of being relieved, were ordered to remain on deck.

Then the gear of the mainsail, tacks and sheets were got clear for running, and in the meantime the darkness had become almost impenetrable, being made visible by the phosphorescent glare of the breaking seas; while the shrieking of the gale and roar of the sea rendered orders all but inaudible. All being ready, the men jumped forward and clapped on to the jib-downhaul, and simultaneously the helm was put "hard-up" and the ship swung off before the wind. The jib-halliards were then let go, and the sails hauled down, when the watch jumped out on the boom to stow it, and each time the ship dipped into the sea they were covered with a ton or so of green water. Meanwhile, the gear of the mainsail was manned, and the sail was run up snug; and then the men clapped on the main-braces and ran the yard in square, hauling the braces well taut. Then the main topsail halliards were let go, and reef tackles manned. Both watches now laid aloft and close-reefed the main topsail, and then laid down on the main-yard and reefed the mainsail, then furling it, and thence down on deck to haul taut the halliards. From here they jumped forward and hauled up the foresail, and then laid aloft again to close-reef the foretop sail, reefing the course as they came down; and then the mizen-top-sail was clewed up and furled.

The ship being now as snug as she could be made, the watch was sent below with orders to be ready on the instant for another call. They had hardly become snug in their bunks, when "all hands" were called again; and the watch arrived

on deck just in time to realize that the squall had done their
work for them, having in its fury cleaned every inch of canvas
from off the yards. Nothing further could, of course, be done
but to goose-wing the main trysail and heave her to until
daylight.

When one realizes the actual condition of affairs—a
hurricane blowing, a mountainous sea running, pitchy
darkness enveloping everything—some concepton may
be formed of the consummate skill and coolness needed
to meet the emergency, and of the qualities essential to
the successful command of a crack sailing packet.

The *New World* was later commanded by Captain
Hale Knight, who was so popular with his passengers
socially that he was nicknamed the "Chivalrous Knight."
Some years afterwards James H. Chamberlain, who had
shipped as a sailor before the mast under Captain Knight,
succeeded to the Captaincy and sailed her many years.

This gallant old packet continued crossing the Atlantic
for many years, under various commanders, among them
being Captains Arthur Champion and William H.
Hammond. Grinnell, Minturn & Co., operated sailing
vessels between New York and the British capital, as
well as to Liverpool until the early eighties, but during
the last ten years or so their ships carried no passengers.

On January 19th, 1882, the *New World* was sold in
London to Austrians and renamed *Rudolph Kaiser*. The
last voyage reported was from Ship Island (Pascagoula)

SWALLOW TAIL LINE PACKET "NEW WORLD" AT LIVERPOOL IN HER PALMY DAYS

She was long unsurpassed in size or popularity amongst the New York-Liverpool Packet Fleet. From original photograph belonging to Mrs. Dora E. Chamberlain, whose husband Captain James H. Chamberlain long served aboard this ship. Perhaps the first photograph of a sailing vessel ever taken.

ALONG SOUTH STREET, NEW YORK, IN PACKET SHIP DAYS

Opposite Grinnell, Minturn & Co.'s Swallow Tail Line Pier in the forties.

July 15th to Sunderland, where she arrived September 10th, 1884.

American packet ships like the *New World* were the master shuttles in the loom of modern civilization, and her extraordinary service of thirty-six years in the transAtlantic trade, commends beyond words the master mechanic who wrought of wood so staunch a craft!

CHAPTER VII

EARLY in 1847, Donald McKay was commissioned by Messrs. Zerega & Co. of New York, prominent shipowners and merchants, who conducted the "Z" line of sailing packets between Liverpool and New York, and were largely interested in supplying the English market with American cotton, to construct the ship *A.Z.* of 700 tons. This vessel proved so successful in quickly carrying her freight and passengers across the Atlantic, that in December 1848, he launched another packet for Zerega & Co.'s account, briefly named *L.Z.*, registered at 897 tons.

It was when the *L.Z.* was on the stocks, in the early part of December 1848, that Donald McKay, deeply engrossed in his increasing' shipbuilding operations, received a blow that staggered him. His wife, Albenia Boole McKay, died after a brief illness. Notwithstanding the numerous cares of a growing young family, she had continually assisted him in his work. Night after

56

TELEGRAPH PRICES AND REGULATIONS.

From BOSTON to NEW-YORK, for first 10 words or less of each message, 50 cents; for each additional word, 3 cents.

From BOSTON to any intermediate Station short of New-York, or between two intermediate Stations, for first 10 words, 25 cents; for each added word, 2 cents.

From BOSTON to PORTLAND, for first 10 words or less, 25 cts.; each added word, 2 cts.

From BOSTON to any intermediate Station short of Portland, or between two intermediate Stations, 15 cents; for each added word, 2 cents.

From BOSTON to NORWICH, NEW-BEDFORD, PROVIDENCE, &c., for first 10 words, or less, 25 cents; for each added word, 2 cents.

From BOSTON to NEWBURYPORT, for first 10 words, 15 cts; for each added word, 1 ct.

From BOSTON to SALEM, for first ten words, 10 cents; for each added word, 1 cent.

Figures are counted as if written into words. The address and signature are not counted. Each three letters or figures of messages written in cipher, are counted as one word, in consequence of the two-fold writing they require.

The original of messages to be sent, must be written plainly, to guard against mistakes.

Each message will be transmitted in the exact order of its reception, first come being first sent—excepting only stipulations in favor of the public press, and the public police.

The Proprietors assume no responsibility in business transacted over their lines, beyond the exercise of good faith, and due diligence, and the amount paid for a transmission.

☞ All Telegraphic Messages are treated as strictly confidential.

Magnetic Telegraph Office,

No. 2 Massachusetts Block, Court-Square—rear of City-Hall.

Boston, ~~Nov~~ Dec 11, 1848.

The following Communication has been received at this Office, by Telegraph, from N. York.

For D McKay

Ship builder, E. Boston

When will ship "L. Z." be launched? Preparing to go Boston. Answer immediately

A. Zerega

11 wds.

COPY OF OLD TELEGRAPH DISPATCH, 1848
Probably the oldest telegram in existence.
From the McKay Collection.

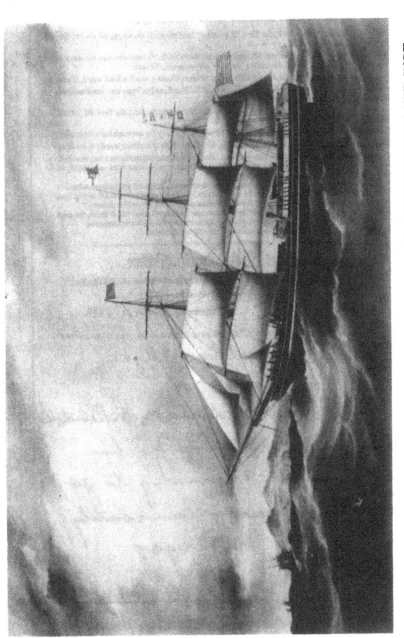

"ANTARCTIC," ONE OF ZEREGA & COMPANY'S "Z" LINE PACKETS, PLYING BETWEEN NEW YORK
AND LIVERPOOL

From the original picture in the MacPherson Collection

night, together they toiled, planning and designing sailing craft. Her father and brothers were actively engaged as shipbuilders in New York, where the sharp-built clipper model now was the sentient thing, so she possessed intimate knowledge relative to their design and construction. It followed, naturally, that long before the people of New England were conversant with this wonderful innovation in marine architecture, the McKays had completed plans for a clipper ship. Boston merchants, for a while could not be persuaded to adopt the clipperly model, then looked upon solely as a New York production, so the introduction of the clipper ship in New England had to be held in abeyance. It was not until 1850, when the *Stag Hound* was built that Donald McKay had opportunity to make practical use of the knowledge he had acquired in designing the type of craft which brought him fame and fortune.

Mr. August Zerega, the head of this once well-known and prosperous concern, was certainly an odd commercial genius. Transacting an immense business that ran into thousands of dollars, he kept no account books, bills were immediately paid in cash and this honest old merchant depended entirely upon a prodigious memory in all his numerous business transactions. It is said that he came on to East Boston, at stated intervals, with currency, and personally paid Mr. McKay as his ships were in various stages of construction. He showed a confidence in his fellowmen that our present-day merchants would do well to emulate, but the absence of recording business transactions, bookkeeping, etc., would seriously cripple any business concern nowadays.

As an example of incoming freight shipments on a Western Ocean packet may now be interesting, the following cargo arrived at New York from Liverpool aboard the *A.Z.*, Captain Chandler, in May, 1850:

Coal, 151 tons, consigned to Manhattan Gas Co.
Rail Road Iron, 930 bars consigned to W. F. Weld & Co.
Merchandise, 23 bales consigned to Cameron & Brand
Merchandise, 3 cases, consigned to A. T. Stewart & Co.
60 casks, consigned to G. B. Morewood & Co.
1 cask Earthenware, consigned to G. B. Morewood & Co.
23 crates, consigned to J. Whittaker & Son
6636 bars Iron consigned to Faber & Bierwirth
150 tons Pig Iron, consigned to Faber & Bierwirth
176 crates Earthen ware, 7 casks, skins, etc.
17 crates Earthenware, 2 bales—all consigned to "Order"
Sundry packages to Various persons.

From the foregoing we surmise that English coal was used to make gas for illuminating purposes in New York; the Railroad Iron (undoubtedly rails) found its way upon pioneer American railways; and the bars of Iron and tons of Pig Iron, were then needed by our manufacturers because this country's immense iron production had not then been developed sufficiently.

In 1848, seamens' wages with small stores, according to the N. Y. Shipping and Commercial List, were:

To—Liverpool and Havre	$15.	per month
North of Europe	15.	per month
Mediterranean	15.	per month
East Indies	12.	per month
South America	13. @ 14.	per month
West Indies	15.	per month
Coasting	16. @ 18.	per month

For the most part, thanks to the attention of the pawn-broker, rum-seller, and boarding-master, the crews of these and other American packet ships were unimpeded with baggage; not a few faced the winds and waves of the North Atlantic clothed only in thin dungarees, facetiously styling their lack of raiment as "flying light." Although for quickness in making sail, reefing or unfurling, the swaggering, roystering "Western Ocean laborers," or "packet rats" as they were called, had few equals, yet the more scientific part of a seaman's avocation was seldom learned, or if learned, very soon forgotten through want of practice. Numerically, packet crews were of the smallest, compatible with the work to be performed; hence it was a constant and terrible grind on the part of officers and men, and the latter were only held subject to an iron discipline by the liberal use of the belaying pin.

As a good indication that his services had been appreciated, Mr. McKay built a third ship for Zerega's Packet

Line, the *Antarctic*. She registered 1116 tons, was designed to sail fast and stow a large cargo. Launched in September, 1850, she shortly afterwards, under command of Captain Ricker, who previously commanded the *A.Z.*,—went through one of the worst winters in the history of North Atlantic shipping—that of 1852-3. Homeward bound on her maiden voyage she was obliged to put in at Hampton Roads after a very severe passage from Liverpool, with the added horror of sixty deaths in the ship from smallpox. The toll of deaths on the Atlantic that winter was indeed a severe one, and almost entirely due to the bad weather.

During the year 1856, Donald McKay constructed the packets *Adriatic* and *Baltic* for Zerega & Co. The latter, registering 1372 tons, was launched in October, 1856.

Most of his thoughts were of the world of waters and the best way to meet its ever-varying dangers, so Donald McKay's financial misfortunes, during the autumn of 1856, deserve only slight mention in these pages. When he failed, Mr. McKay was in hopes that his assets would pay all his debts, and leave him something upon which to begin the world again, but the extreme depression of business, and consequent fall in the value of real estate, in which his money was invested, convinced his creditors that to force a sale would only ruin him, without benefiting them. Encouraged by Enoch Train, his first and best friend, and by many others, he eventually resumed his career of usefulness. The McKay shipyard lay idle for a time, as shipping business was almost at a standstill. The year 1857 saw many ships tied up in American ports while few at sea could clear expenses.

CHAPTER VIII

ONE Sunday morning in January, 1847, when Boston's streets had been completely covered with several inches of snow, a sturdy, well built man in his thirties, accompanied by a likely-looking lad of twelve years, happy in the possession of a new bright red sled, walked off one of the East Boston ferryboats. They had not proceeded more than a couple of blocks, when the man was attracted by two men engaged in animated conversation before a shoe store. From their garb, and the tell-tale gnarled hands, one easily recognized them as shipyard workers, and their discussion had assumed that serious stage where finance played an all-important part. Shoes—boots rather—they both needed badly for the morrow, but their combined funds would not permit any extravagance whatsoever, and a good serviceable pair of "high tops" attractively displayed in the show window, cost a trifle more than either man's repeated counting of his funds permitted. As one of them disconsolately turned away from viewing the much coveted footwear he recognized the man drawing the sled as his employer, Donald McKay, shipbuilder of East Boston. The boy, too, they knew well, for little Cornelius was constantly around his father's shipyard and a prime favorite with the workmen.

Mr. McKay, took in the situation at a glance. Work had not been plentiful of late; true he had only a short time previously finished construction of the largest merchant vessel afloat—the three-decked packet *New World*, but now there was no ship on the stocks in his yard. These men, both skilled shipwrights, were working only on part time, employed in various old-fashioned ways that then prevailed when work was slack or weather conditions too bad, and a shipbuilder desired to retain his high class labor, such as making treenails in the shed out of sticks of wood with axes, etc. After a short conversation, the shipbuilder invited both men into the shoe-shop with him. Once inside, brief orders were given by Mr. McKay and the storekeeper soon outfitted them with good serviceable leather boots. A few minutes afterwards, the workmen, each with a package under his arm, were profusely thanking their benefactor.

Continuing on his way, Donald McKay, and Cornelius reached the Common. At that time this was really Boston's principal playground. Here both joined the happy throng of children and grown-ups. Remembering his own boyhood days spent amidst Nova Scotian winters, the young shipbuilder greatly enjoyed speeding across the snow-clad meadows of the Common or along its trodden paths, with his little son holding fast to the sled.

After indulging in this healthful exercise a while, he drew out a handsome bull's eye watch which had been given to him by his staunch admirer, Dennis Condry. This indicated it was time to leave the Common, so father and son proceeded to the beautiful home of Boston's famous shipping merchant, Enoch Train, located nearby, at 70 Mt. Vernon Street, Mr. Train opened the door himself and ushered them into his library. Here, seated at a large mahogany table, covered with ship plans and numerous drawings, they found another of Boston's leading merchants, Robert G. Shaw, who gave them a cordial greeting.

A long, earnest discussion took place between the two merchants and the shipbuilder, culminating in a contract for construction of the largest vessel ever built up to that time for Train & Company's Liverpool packet service. This was the *Ocean Monarch* of 1301 tons register. She was launched some five months afterwards, in July, 1847.

How rapid had been the growth of Train's Boston and Liverpool shipping is evidenced by the enlargement of their packet ships, within less than three years, from 620 tons as represented by the *Joshua Bates*, up to their present construction of what was then regarded as a

ENOCH TRAIN

Well known merchant and shipowner of Boston.

A VIEW OF BOSTON HARBOR IN 1848 AS SEEN FROM EAST BOSTON

Shows Train and Company's Wharf in the distance, while some of their Packet Ships can be seen at anchor, and one of them, to the extreme right of picture is sailing gallantly under her own canvas.

Reproduced through the courtesy of Massachusetts Historical Society.

veritable monster of the sea. Due to the wonderful growth of this country and Canada, the trans-Atlantic packet business increased rapidly; and the Irish famine of 1846, gave great impetus to immigration here.

The *Ocean Monarch* was placed under command of Captain Murdoch, who had previously commanded the *Joshua Bates*, a deservedly popular Yankee Skipper, considered by the owners one of their most capable shipmasters. He sailed her during the ship's somewhat checkered career, making one passage from Boston to Liverpool in 14 days and some hours, although becalmed about two days.

Her end came on August 24th, 1848. When about eight hours from her dock at Liverpool, this fine ship was completely destroyed by fire, with the loss of about 400 lives. The following vivid account is taken from George Francis Train's biography *My Life in Many States:*

In '48 I was at the pier one day on the lookout for the *Ocean Monarch*. Although the telegraph had been established in '44, it had not been brought from Nova Scotia to Boston, and we had only the semaphore to use for signaling. When a ship entered the harbor, the captain would take a speaking-trumpet and, standing on the bridge shout out the most interesting or important tidings so that the news would get into the city before the ship was docked. The *Persia* was also due with Captain Judkins and it came in ahead of the *Ocean Monarch*. Some three or four thousand persons were on the pier waiting eagerly for the captain's news. I was at the end of the pier, and saw Captain Judkins place the trumpet to his lips, and heard him shout the tidings. And this is what I heard:

"The *Ocean Monarch* was burned off Orm's Head. Four hundred passengers burned or drowned. Captain Murdoch taken off of a spar by Tom Littledale's yacht. A steamer going to Ireland passed by, and refused to offer assistance. Complete wreck, and complete loss."

The captain shouted hoarsely, like a sentence of doom from the last "Trump." Every one was stunned. The scene was indescribable both the dead silence with which the dreadful tidings were received, and the wild excitement that soon burst forth.

CHAPTER IX

McKAY packet ships were celebrated for their strength; they were designed to carry a tremendous press of sail in heavy weather without straining. In light winds they were not so fast, but the North Atlantic packets did not sail in the latitudes of light winds. By last accounts received from Liverpool just previous to this packet's launch, three ships constructed by Donald McKay had made the shortest passages to that port from the United States among a large fleet of fast sailing vessels. The *John R. Skiddy* landed her passengers at Liverpool in 14 days and a few hours after taking them on board at New York. The *Ocean Monarch*, from Boston, made the passage in nearly the same time, although she was becalmed about two days, and the *New World* went out in 17 days from New York. These results spoke for themselves.

Here was another vessel to fly Train & Company's signal, a red field with white diamond in centre; and, by the black "T" in their foretopsail below the close reef band, they were easily recognized at sea.

The *Anglo-American* was strongly built in the best manner and beautifully modelled. The arrangements of her deck resembled the *Washington Irving*, viz.: she had a full poop, a topgallant forecastle, a large house amidships before the main hatchway, and a deep waist. Her length of keel was 145 feet, on deck she was 150 feet and over all 156; breadth of beam 33 feet, depth of hold 20 feet. The bow was very sharp and most beautifully formed. It had less flare than the bow of any other Train packet and consequently the sweep of her model at the rail was more easy in its curvature and fairly clipperlike in its termination. Her lines were truly rounded, notwithstanding the sharpness of her ends, and her sides smooth and regular as those of a pilot boat. She was built of the best seasoned white oak, hackmatack and yellow pine, than which there was no better shipbuilding material obtainable in this country.

The workmanship and designs of her cabins were considered superior to those of any packet plying between Europe and America. For the stowage of cargo or the accommodation of steerage passengers, her between-decks were admirably adapted.

Under command of Captain Albert H. Brown, the *Anglo-American* sailed from Boston to Liverpool on January 5th, 1848, and made a right smart passage, considering the season of the year.

At the end of 1847 the Managers of the Cunard Line

THE WHITE DIAMOND LINER "ANGLO AMERICAN"

One of Train & Co.'s famous Boston-Liverpool packet ships that so long and successfully maintained trans-Atlantic service between New England and Great Britain. From a picture in the McKay Collection.

found it necessary to abandon their purpose of making Boston their sole American port, and began to send half of their ships to New York. Previously the British Government required a double service and increased the Cunard compensation to £173,340 per annum. To comply with this new requirement four (4) new steamships (all side wheelers) were built. Messrs. Train & Co. and the various New York packet ship owners are certainly entitled to great praise for the energy, skill and perseverance which they continued to display in the organization and maintenance of their splendid lines of packets, especially in view of the indifferent attitude assumed by the United States Government.

If assistance had then been rendered American ship-owners in the trans-Atlantic trade, Great Britain would not have enjoyed that ocean-carrying supremacy which placed her in the forefront as a maritime power; and American shipping would not have been relegated to an obscure place in international commerce previous to the outbreak of our Civil War.

During this,—the Packet Ship Era—American packets, and transient ships too, surpassed any craft afloat. The flag of this country predominated in the docks of Liverpool and other ports of Europe. American ships furnished both London and Liverpool with food supplies, as well as cotton, and other staple products for Britain's factories. Our trans-Atlantic packets enjoyed a monopoly of the

European emigrant trade; Train & Co. being in especial favor with the Irish who embarked in large numbers for America upon their ships. It was not until the latter part of 1852 that the Cunard Co. dared to fit their vessels with accommodations for emigrants. In such high favor were American vessels held that Bremen and Hamburg drew their supply from us for some years.

Captain Brown continued in command of the *Anglo-American* until December, 1849, when he took over Train & Company's ship *Parliament* directly after she was launched from McKay's yard.

En route from Boston, November 7th, 1850, in a sudden shift of wind, the *Anglo-American* carried away her main topmast by the cap, while six men were furling topgallantsail, all of whom were drowned. The weather was so severe she could not get any canvas above a reefed mainsail for eight days; two days after, split foresail from foot. Although in a crippled state, she finally made port without further casualties.

Under various commanders this ship continued to enjoy wide popularity as she plied the Atlantic between Boston and Liverpool, until about 1852, when she was sold by Train & Co. and went under the British Flag.

To quote George Francis Train, in a letter from Sydney, N. S. W., dated March 7th, 1854, the career of the *Anglo-American*, as such, was avowedly ended.

But what American ships are those under the stern of the *Golden Age,* bearing the English Flag? 'Pon my soul, they are the old Boston packet ships *Washington Irving* and *Anglo-American,* the pioneers of Train's line, that were sold in London some two years since! I saw them in Melbourne when I arrived last May; now here they are at Sydney, they must have made quick trips to England. Although you have built the *Sovereign of the Seas* and the *Great Republic,* you have no reason to be ashamed of your old acquaintances, Mr. McKay. The *Washington Irving* still holds the name, but the *Anglo-American* flourishes under the more supercilious soubriquet of *Arrogant.*

CHAPTER X

WHEN Donald McKay kept producing many of the best, largest and finest trading vessels and packets in New England, he achieved success in the construction of a class of ships that differentiated from other merchant sailing craft—the Cotton Ships or "Kettle Bottoms" as they were called.

The *Jenny Lind* was his first contribution to the cotton-carrying fleet. She was owned by Messrs. Fairbanks & Wheeler, well-known merchants of Boston, who were engaged in supplying the Liverpool market, as well as New England factories, with American cotton, consequently this ship sailed in both the trans-Atlantic and American coast-wise trade out of Boston.

She was often engaged in making the "Triangular Run"—Boston to New Orleans, New Orleans to Liverpool, and Liverpool back to Boston. As the European cotton-carrying trade did not flourish all year round, these vessels were well fitted for passenger traffic. The return passage, Liverpool to this country, proved a highly successful venture if the ship had her cabins filled with passengers and a goodly number of emigrants aboard, together with cargo which generally consisted of salt, iron and general merchandise.

Upon one of her early voyages, homeward bound from Apalachicola, Florida, with 1885 bales of cotton, the *Jenny Lind* went ashore at Cohasset, Mass., February 18th, 1849. She was commanded by Captain Bragdon, who soon landed most of her cotton cargo, and as she had gone on a smooth beach, the vessel suffered little damage.

Later on this cotton carrier made some excellent passages across the Atlantic to Liverpool, under command of Captain Lauchlan McKay, one of Donald McKay's brothers. It was his able management of the *Jenny Lind* that probably influenced Donald McKay to place him in charge of the *Sovereign of the Seas*.

COTTON SHIPS AND THEIR CREWS, CHANTEY-MEN AND CHANTIES

It must be borne in mind that cotton ships like the *Jenny Lind* were a special class. Obtaining fifty or one hundred per cent more capacity than the registered tonnage, was the common practice adopted in these vessels. Great capacity was the prime consideration. Before the powerful compressing apparatus of today was invented a gross ton of cotton occupied about 100 cubic feet of space, the reduction to 60 cubic feet being a later achievement.

To carry a good cargo of this bulky commodity, therefore, required a large vessel; and as a large ship is expensive, owners prudently studied how to gain the capacity they wanted without at the same time incurring new cost of operation. Tonnage taxes and port charges were the expenses most dreaded, as being the least within the control of owner or master after the vessel was launched.

Shipbuilders had to take advantage of the laws for the measurement of vessels to reduce the official tonnage of the cotton ships considerably. Thus the common practice of obtaining much more capacity than the registered tonnage was adopted. The bow was lengthened out under water, so that the deduction of three-fifths of the beam made the tonnage length less than the actual length. They also made the hold much wider at the water than on deck, and much deeper than one-half the beam, and constructed large poop and topgallant-forecastle decks, covering nearly the whole top of the ship. These were open, however, amidships, and good for the stowage of 200 or 300 bales of cotton, which escaped tonnage taxation. The result was a roomy ship, with the old-fashioned tumbling home of the top sides, which passed muster at the custom-house as of far less capacity than she really had. The government did not get its just dues, but shipping was benefited.

As no roomy ship could ever be loaded down to her deepest draught with so light and fleecy a cargo as cotton,

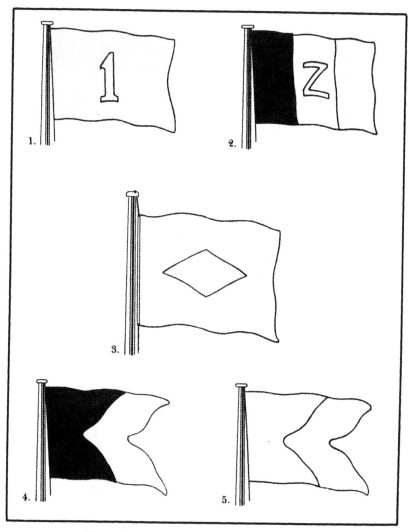

House Flags Carried by Donald McKay's Packets.

1. Andrew Foster & Son, of New York. *White number 1 on red* field.*
2. Zerega & Co. of New York "Z" Line. From left to right: *Blue stripe, red Z on white stripe, red stripe.*
3. Enoch Train & Co., White Diamond Line, Boston–Liverpool Packets. *White diamond on red field.*
4. Grinnell, Minturn & Co., of N.Y., Liverpool Line. *Blue stripe, white stripe.*
5. Grinnell, Minturn & Co., of N.Y., London Line. *Red Stripe, white stripe.*

*In the original edition, where the plates were in color and bore no verbal descriptions, the red used was of a very orange quality. Unsure of what was intended, we say "red" here for brevity.

it was customary to stow away bales in every available sheltered space to be found on board, from the limber strake to the main rail, and even the mess-table in the cabin was often a bale of cotton. No matter how big the cargo, the ship would not be down in the water to her bearings, and would be top-heavy and crank in consequence; so that it was always necessary to carry from 100 to 300 tons of stone ballast for the sake of stability, and even then the "kettle bottoms" were apt to go away over on their sides whenever the wind was abeam, and stay there, to the discomfort of all on board. The sailor loves to see a good space between the deck he treads and the water upon which he floats; and whatever beauties the owner saw in a "kettle bottom" that carried a big freight and paid him well, Jack saw none. His preference has always been for a ship that would stand up stiff under sail, and few of the cotton fleet would.

American cotton ships by reason of their superiority and the able management of them in time began to do about two-thirds of the business there was in their line from all ports of the United States, and thronged the wharves of New Orleans, Mobile, Savannah, Charleston and Apalachicola, also New York and Boston.

There was a hard class of men, called "hoosiers," who worked in New Orleans, Mobile and other Southern ports during the winter stowing these ships with cotton, and in the summer sailed aboard the Atlantic packet ships.

They were all good chanteymen, who would constantly sing at their work and were good-natured, hard workers, but they could not be brought under strict discipline by the ship's officers. To them is attributed the introduction of chanteying aboard American ships, which not only added cheer and pleasantness to sea life, but was of material assistance in getting a crew to willingly pull and haul as well as do other laborious work requiring concerted action.

Aboard a sailing ship it is an inestimable asset, for chantey singing produces team work and co-ordination. A good chanteyman was soon an acknowledged leader. On the forecastle head, heaving on the windlassbrakes or walking 'round the capstan, pushing on a capstan bar, the chanteyman sings the solo while his mates join in the refrain. Popular chanties aboard American packets, clippers and other craft were:

"Blow, my Bully Boys, Blow," "Shenandoah," "Rio Grande," "Yankee John Storm Along," "The Capstan Bar," and last but not least those innumerable verses to that much-courted lady "Sally Brown."

To masthead a topsail yard, or when doing any hauling of length, a *long drag chantey* was sung, such as "Reuben Ranzo," "Blow the Man Down," "Whiskey for my Johnnie," and that "hoosier" favorite "Roll the Cotton Down."

Hauling on a rope already taut, or to sweat up on a

halyard, oldtime sailors sang what was styled a *short drag chantey*—"Haul the Bowline" and "Paddy Doyle."

There were many other chanties besides those above mentioned and they varied somewhat in tune; furthermore a clever chanteyman improvised as he went along, so the rendition of these sea songs differed continually. In rhyme he sang of the virtues and failings of the Captain and the crew or a host of others.

Today the chanteyman is veritably a "rara avis"— his usefulness belied and belittled in every way by changed conditions, aided by modern mechanical appliances. In his virtual extinction however, he carried with him the beloved ship of sail that he sang in.

> But the advent of Steam and the eight-hour day,
> With Union rules and increased pay,
>   But little romance leaves
> Of the days when the Chanteyman led us in song,
> In "Blow the Man Down" and "Old Storm Along";
> When we stood by the braces, ready for the call,
> And the yards swung round to "Mainsail haul,"
>   To the music of rattling sheaves.

CHAPTER XI

HERE is a vessel that Boston's leading ship-owner and merchant venturer, who had deservedly become lord of the ocean pathway connecting Great Britain and New England added to his fleet.

Changed and varying conditions in Europe gave a great impetus to emigration and the "Packet Line" business was very brisk, continuing so for a number of years. Those coming here, when able, sent funds and passage tickets to relatives, as well as friends left behind in Europe and it is stated, on good authority, that Enoch Train & Co., who specialized in small exchange drafts for one pound or so, averaged $1,000,000 in receipts each year; very large indeed for those days and a great help to oversea passenger business. Cargoes of freight on both sides of the Atlantic, as well as passenger traffic, kept increasing, as commercial relations grew closer between the Old and the New World, consequently instead of despatching a ship from Boston, say in six weeks, regular weekly sailings from Liverpool were maintained and semi-monthly from Boston. Fair weather or foul they sailed as advertised.

Although Train's White Diamond fleet was being steadily augmented, the loss of two vessels the *Anglo-Saxon* and *Ocean Monarch*, aggregating nearly 2200 tons, was no inconsiderable factor toward causing them to charter ships; and they did this while some of them were upon the stocks in the McKay shipyard. There was now a great demand for ships—the clipper era was dawning and a complete reversal in marine architecture was well upon its way.

The *Plymouth Rock* though originally designed for the freighting business, was, nevertheless, arranged in all her appointments equal to any packet ship of her size. Messrs. Train & Co. perceived this, and at once chartered her from George B. Upton and Capt. Eben Caldwell for whose account Mr. McKay constructed her. Registering nearly a thousand tons, her between-decks lofty, well-lighted and ventilated, and her arrangements on deck all that skill and experience could produce, she was one of the best adapted passenger ships afloat.

Under command of Capt. Caldwell, she took her place in Train & Co.'s line, and sailed for Liverpool, March 5th, 1849, the day of Zachary Taylor's inauguration, making a very good passage across the Atlantic considering the time of year.

From an old packet captain's *Reminiscences* we supply the following:

Handling sailors in those days was a problem, but handling several hundred emigrants in the winter months was equally bad. These latter were the rakings and scrapings of all Europe. Men, women, and children tumbled into the between-decks together, dirty, saucy, and ignorant, breeding the most loathsome of creeping things. The stench below decks, aggravated by sea-sickness and the ship's poor equipment for the work, placed us far below the civilization of the Dark Ages. It was not uncommon in midwinter to be fifty or sixty days making the homeward passage. In gales, which were frequent, hatches had to be battened down, and men, women and children screamed all night in terror. Ship fever, small-pox, and other contagious diseases were common, and it is a wonder that as many survived the voyage as really did. Rations were served out once a week in accordance with the British Government allowance—just enough to keep starvation away. The estimated cost to feed them was twenty cents a head per day. The steerage passage rates were four pounds, and between passenger and freight money, the ships generally paid very good dividends.

The *Plymouth Rock* remained in the North Atlantic packet service, under charter to Enoch Train & Co. until about 1852.

It was on this vessel that George Francis Train sailed away from Boston when the Australian gold fever was at its height to organize the house of Caldwell, Train & Co. in Melbourne, Australia. Everything was taken from Boston — clerks, sets of books and business forms.

Nothing was left to the chance of getting these things in Australia. This commercial venture proved very successful; and it was singular that a vessel with the historic name of *Plymouth Rock* should have been chosen to bear such an enterprising Yankee expedition into the South Seas.

She then sailed on the Liverpool-Melbourne run during the gold rush to Australia, afterwards engaging in various trades under different owners.

It is related that in the seventies this vessel left Gravesend with 20 foremast hands, not one of whom could steer and but very few of them had ever been to sea before. Forty-two days later, when the *Plymouth Rock* reached New York, her mate had turned this gang of hoodlums into very fair seamen, all of whom steered sufficiently well to satisfy his critical eye.

CHAPTER XII

LAUNCHED JUNE, 1849

W E are now in the midst of what has been called "Argonaut Year"—1849, and the ship *Reindeer* of 800 tons has just left the ways at the McKay yard. Owned by George B. Upton, one of Boston's most enterprising merchants, she is going to be loaded quickly for California.

In September, 1848, news reached Boston of the discovery of gold in California. During the next three years California was transformed from a wild region, containing about 15,000 white population into a State, with more than a quarter of a million people. Of course, in this new settlement materials required for necessity and comfort were demanded, and Boston was in the front of this enterprise, sending such merchandise as might be thought fit for a market.

Mercantile adventure with California was most uncertain, owing to the infrequent and slow communication and consequent lack of sufficient information. With the whole commercial world seeking the new market and no advices as to stocks of merchandise on hand in San Francisco shipments were in many cases pure speculation; sometimes resulting in enormous profit.

Merely by way of illustration we may relate the amusing tale of one Boston shipper who, from such calculation or guesswork as circumstances allowed, concluded that some commodity—we believe it was flour—would be in demand when the ship *Reindeer* should reach San Francisco and loaded for that port accordingly. When loading, he happened to find obtainable a great quantity of damaged dry beans, which he purchased for a trifle. On arrival at San Francisco the harbor was full of newly arrived flour, for which no price could be obtained. But there were no beans to be had, with a keen demand for them, and our shipper realized a handsome profit. Such were the chances of California trading in 1849–50.

Leaving Boston, November 23rd, 1849, under command of Captain Lord, the *Reindeer* arrived at San Francisco, April 2nd, 1850, 130 days afterwards—a slow passage in the light of what McKay's clippers accomplished within the next two or three years. She was the first of his ships to enter the California trade, and he had vainly endeavored to induce Mr. Upton to let him construct a sharp built ship, for he was familiar with John W. Griffith's clipper productions, *Rainbow* and *Sea Witch* and others that had already been produced by New York shipbuilders.

Reindeer's following California voyage, from Norfolk, Va., consumed 148 days.

From San Francisco she sailed to China, thence to Boston. Prof. Morison, in his *Maritime History of Massachusetts,* shows by the following account how a law of 1817, requiring two-thirds of an American crew to be American citizens, was disregarded during the year 1851, upon this vessel when homeward bound.

A sample crew is that of the ship *Reindeer,* Canton to Boston: two Frenchmen, one Portuguese, one Cape Verde Islander, one Azores man, one Italian, one Dutchman, one Mulatto, two Kanakas, one Welshman, one Swede, two Chinese, and two Americans.

On her next passage the *Reindeer* arrived at San Francisco in 132 days, from New York.

On this and two subsequent passages to California, she was commanded by Captain Bunker.

This ship was lost near Manila February 12th, 1855.

CHAPTER XIII

"CORNELIUS GRINNELL," 1118 TONS
LAUNCHED JUNE, 1850

THERE was keen rivalry between the various packet lines plying between New York and Liverpool; Charles H. Marshall's Black Ball Line, Taylor & Morrill's, Collins' Dramatic (precursor to the famous American Line of steamships), Williams & Guion Line, Ogden's and Grinnell, Minturn & Co.'s fleet. The struggle was an animated one, when this last-named firm chose Donald McKay to build a vessel that could successfully compete against the "flash" packets produced by New York shipbuilders. New York and Boston packets surpassed any ships in the world until the California fever broke out, for speed, safety and beauty.

The influence of steam on the Atlantic had just begun to be felt, and was an incentive for increased effort upon the part of sailing vessels to make short passages, when Moses H. Grinnell, contracted with Mr. McKay to build the *Cornelius Grinnell,* for the Liverpool Swallow Tail Line of packets. The success of an individual ship depended then upon her record for speed and regularity.

She was 172 feet long on the keel, 180 feet on deck, had 38 feet extreme breadth of beam and 23½ feet depth of hold, including 7 feet, 10 inches height of between decks. Aloft as well as below, everything was of the best materials, and after being fitted out, she was certainly the strongest ship of her size that had ever been constructed in America.

Notwithstanding all the substance of timber, and strength of fastening, it had been found that most ships in the trans-Atlantic trade, after running a few years, would settle or sag, and to obviate this the *Grinnell* was built so strongly that it was almost impossible for her to hog or sag until her back would have been actually broken.

She was placed under command of Captain William Howland, who had superintended her construction. He was an aristocratic captain, an oddity in the American packet service. He would never allow himself forward of the mainmast and very seldom spoke to a sailor, but gave all his orders to the chief mate. He only came on deck at stated times and always wore kid gloves. He was a good navigator but not much of a sailor, having taken command without going through the forecastle—one of those masters whom sailors facetiously described as being "blown in through the cabin window" instead of "crawling through the hawse pipes and working their way aft."

This packet's rigging was made in New York. Ship-owners of that city then often selected local riggers to

CARGO HOLD OF THE NEW YORK-LIVERPOOL PACKET "CORNELIUS GRINNELL," 1118 TONS

When fitted out in 1850, this ship was said to be "by all odds" the strongest of her size afloat.

From a painting made by Harry Saunders in the McKay Collection.

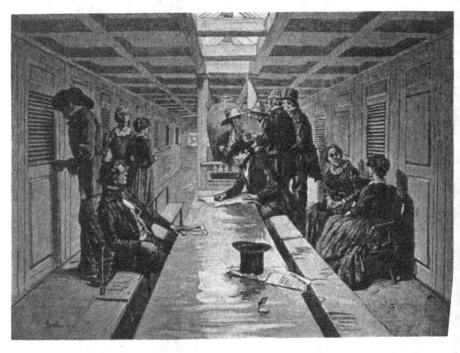

INTERIOR OF THE SALOON OF A SAILING PACKET-SHIP

put the finishing touches aloft. Shortly afterwards she took her place on the berth to Liverpool.

Upon arrival in England this vessel excited much curiosity among persons interested in shipbuilding, and it was conceded she was "by all odds" the strongest merchantman of her size afloat.

A fine sight it was to see a returning packet like the *Grinnell* come up the East River with all sail set and anchor off the Swallow Tail Line Pier, at the foot of Pine Street. News of her arrival had been conveyed by signal-telegraph from Sandy Hook, and thence by another signal-telegraph to New York, though perhaps she had been over forty hours in sailing her last twenty miles from the Hook to the river. It is pleasing to recall, now, that whenever a shipmaster thus brought his vessel in, there were many persons gathered at the pier who greeted his exploit with applause.

Atlantic crossings, of course, were not always fast; the average length of a sailing passage from Liverpool to New York being estimated at about 34 days, and from the latter to the former in about 20 days, the prevailing winds being westerly. One might suppose that the days would seem long and monotonous, but old sea travelers averred they did not find them so, and it was all so strangely fascinating! But passengers must amuse themselves and from an article in *Harper's Magazine*, we culled the following:

After a nine-o'clock breakfast, backgammon, chess, checkers, and shuffle board were in requisition, and if the ship was rolling it required considerable skill to keep the lignum-vitæ blocks in the latter game from sliding. At twelve o'clock the sun was taken and the ship's reckoning made. Betting on the runs was much less frequent than in these days when our huge modern leviathans make the result more uniform.

Dinner was served at half-past two and eaten in about two hours. The bill of fare of a Christmas dinner on board the *Cornelius Grinnell* in mid-ocean on the way to England in 1858, is preserved in some modest verses written at the time by one of her passengers.

First of all we had some soup, and it was very good;
But as I could not take it, I left it for those who could.
The next course was boiled cod-fish, and boiled potatoes too;
But that I do not like, so I left it for those who do.
The next course was a stunner, which I must try to relate,
But I could not get a little of each dish upon my plate.
We had a fine roast turkey, just as fine and good,
As if you had just gone and shot it on the prairie or the wood.
A fine dish of stewed chicken, a fine macaroni pie,
Roast and boiled potatoes, and mashed turnips bye-the-bye.
And very good fresh bread, which the steward bakes each day,
Besides sea-biscuits, pickles, and such fixings in that way.
And when we all had had enough, and that good course was
    done,
On came the fine plum-pudding, and then commenced the fun.
Mr. Clark had brought champagne for himself or for his wife,
And it certainly was some of the best I ever tasted in my life.
He brought it for sea-sickness, but they did not drink it on the
    way.
And he thought we could not do better than drink it on
    Christmas Day.

Tea was taken at seven o'clock, and followed by the reading of the daily or weekly journal of the voyage, and by a lecture from a passenger or by Charades.

Card-playing and singing were favorite amusements of the later hours, few passengers caring to go to bed until mid-night. In this connection, it may not be amiss to say that poker-playing was carried to England in the old packet ships, and many a noble son of Kentucky beguiled the tedium of a long off-shore voyage by teaching John Bull this little game.

For several years the *Cornelius Grinnell* remained one of the most popular packets that rendered illustrious service to the growth of this country. Although she made no extraordinary trips across the Atlantic, this staunch vessel was conspicuous for sailing regularly, and successfully maintained an enviable reputation for arriving safely in port with passengers and her cargo, in good condition.

She was commanded by Captain John W. Burland, from 1874 to 1882, when Grinnell, Minturn & Co., retiring from the shipping business, sold her as well as other vessels of their well-known trans-Atlantic fleet.

Sailing from New York January 11th, 1883, for Antwerp, with cargo of oil, etc., she sprang a leak when off Montauk Point and returned to Whitestone, L. I., thence towed to New York for drydocking. After being overhauled and repaired, this vessel with 340 barrels of petroleum on board, took fire at foot of Court Street, Brooklyn, and was seriously damaged. She was however, sold at auction March 27th, 1883, to C. A. Williams of

New London, Conn., for $4,500 and changed to a coal barge.

Towed astern of an ocean-going tug, at the end of several hundred fathoms of hawser, our former crack packet of the Swallow Tail Line, hogged and razeed, has finally become a coal carrier. Together with a number of other "barges," she is pulled across New York Bay to be unburdened of her cargo of "black diamonds" at some New York or Brooklyn pier. How her timbers must have thrilled as, entering that expanse of the East River between Brooklyn Heights on the one hand and lower South Street on the other, she recalled the days of her glory! What memories must have crowded upon her! And if those who turned idly to watch "that tow" go by, seeing her only as one of several clumsy-looking coal-begrimed barges, had been able to understand the language of ships, what a tale she could have told them —a tale of shores lined with enthusiastic multitudes watching her as she majestically threaded her way up the East River under her own immense mass of snowy white canvas! She could have told them how, with other Yankee packets she had seized a monopoly of the ocean-carrying trade between Europe and America, that contributed immeasurably to the growth and benefit of both continents and their peoples!

CROSSING THE ATLANTIC IN PACKET SHIP DAYS

Swallow Tail Line packet *Cornelius Grinnell* outdistancing a paddle-wheel steamer in mid-ocean. From painting by Fred S. Cozzens, in the author's possession.

CHAPTER XIV

"MAN overboard! Man overboard!" This agonizing cry rent the air at the launching of Train's handsome packet ship, as the "Immortal Daniel" was in the midst of his oration. A good sized crowd had gathered at the McKay yard to hear America's leading orator and now everyone rushed to the water's edge. Here it was discovered that the rescued one, rudely handled with boat hooks, pikepoles, etc., was none other than one of Boston's leading shipping merchants. Daniel Webster finished his oration abruptly and joined the sympathetic crowd around the recently immersed one,—who falling overboard in a generous state of inebriation, was thereby completely sobered and soon able to join, in the mould loft, the invited guests and others who foregathered after this ship had reached her destined element.

In the production of this truly magnificent vessel, Train & Co. had spared no expense, and the result was, that she was without a rival in New England. Larger ships belonged to New York but none of her class was more beautiful, better constructed or more splendidly finished. She was, like her namesake, great!

On the keel she was 173 feet long; and over-all 186; she had 38½ feet extreme breadth of beam, 24 feet depth of hold, 12 inches dead rise at half floor, 9 inches swell, and 2 feet 2 inches sheer. In model she could be classed a fair medium between the extreme sharpness of the clipper and the fullness of the freighter or cotton ship. Her beauty consisted in the harmony of her proportions.

A full figure of the distinguished statesman whose name she bore, ornamented the bow. It was slightly inclined forward, to correspond with the rake of the stem; and was designed to represent him in debate. Thrown lightly over his left shoulder was a mantle, the folds of which were gathered in front and held in the left hand, while in the right hand which hung by his side, was the Constitution. The design and execution were excellent, and the likeness correct. It was painted white, and the trail-boards, which partly formed its pedestal, were beautifully ornamented with gilded carved work. *Daniel Webster*, in gilded letters, also ornamented the head boards on both sides.

On her maiden trip the *Daniel Webster*, under command of Capt. William H. Howard, passed Cape Clear 13 days, 10 hours from Boston. Beating windward in a hard wind

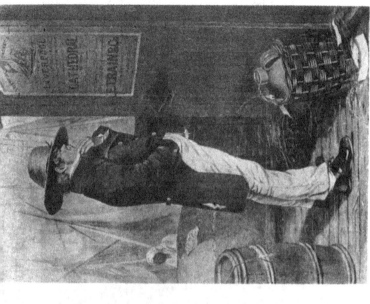

OUTWARD BOUND FROM LIVERPOOL

Showing an Irishman in straitened circumstances examining one of Enoch Train & Co.'s shipping announcements of sailing to Boston on the well-known *Daniel Webster*.

From an old print, Collection State Street Trust Co.

HOMEWARD BOUND FROM BOSTON

Showing the same individual, having become prosperous in the United States, scanning another Train poster giving the next sailing to the "Old Country."

From an old print, Collection State Street Trust Co.

"DANIEL WEBSTER" RESCUING THE PASSENGERS OF THE SHIP "UNICORN" ON NOVEMBER 9TH, 1851.

Reproduced through kindness of State Street Trust Co., Boston.

showed her at her best; indeed this might be said to be the case with all the "White Diamond" packets. Owing to the beautiful flat set of their cotton canvas, and the sharpness with which their yards could be braced up, they could generally head a point higher than other ships.

An interesting rescue of the passengers from the immigrant ship *Unicorn* by the *Daniel Webster* under Captain Howard, on November 9th, 1851, is rather boastfully told by George Francis Train, in his autobiography, *My Life in Many States*. Mr. Train was en route from Boston to Liverpool to assume charge of Train & Company's branch there.

The trip was destined to be eventful. Five days after leaving Boston we ran into a heavy gale from the west. Our boat was sturdy and we had no fears, but I knew that many smaller and less seaworthy ships would suffer in such a driving storm. We were, therefore, on the lookout for vessels in distress.

For the greater part of the time, during the height of the gale, I stood on the bridge closely scanning the horizon line in front. Suddenly something seemed to rise and assume form out of the storm wrack, and this gradually grew into the shape of a vessel. I saw that it was a wreck, shouted to the captain, but he, looking in the direction, could make out nothing. My eyes seemed to be better than his, although his had been trained by long practice at sea. He could not see much better when he got his glasses turned in the direction I indicated, but finally he discovered the vessel, though he did not seem desirous of leaving his present course to offer assistance.

I insisted that we should go to the rescue of the ship and her crew, and he turned and said: "Mr. Train, we sea captains are prevented from going to the rescue of vessels, or from leaving our course, by the insurance companies. We should forfeit our policy in the event of being lost or damaged."

"Let me decide that," said I. "We can not do otherwise than go to the assistance of these persons." And we went. The *Webster* bore swiftly down upon the wreck, which proved to be in worse plight than I had imagined. She was buffeted about by the waves, and seemed in peril of going down at any moment. Men and women were clinging to her rigging, hanging over her sides, and trying to get spars and timbers on which to entrust themselves to the sea. The doomed vessel was the *Unicorn*, from an Irish port, bound for St. John's, N.B., with passengers and railway iron. This iron had been the cause of the wreck, for in the rough weather it had broken away from its fastenings, or "shipped," as the sailors express it, and had broken holes in the sides of the boat and overweighted it on one side.

A brig that had sighted the *Unicorn* before we came up had taken off a few of the passengers—as many as it could accommodate. The *Unicorn* was a small vessel, and there seemed little chance for the rest of the passengers unless we could reach them. The sea was running very swift and high, and it was not possible to bring the *Webster* close to the side of the *Unicorn*. To make matters worse, the sailors had found that there was whisky in the cargo, and in their desperation, drank it without restraint. They were, consequently, unmanageable. They could not help us to assist the miserable passengers on their own boat.

There was nothing else to be done except to get into our small boats and try to save as many passengers as possible. The captain got into one boat and I into another, and we were rowed to the side of the *Unicorn*. There we discovered that many had already perished. Dead bodies were floating in the sea about the ship. We tried to get up close enough to reach the passengers, but found it impossible.

"Throw the passengers into the sea," I shouted to the captain of the *Unicorn*, "and we will pick them up. We can't get up to you." In this way the crew of the *Unicorn* throwing men and women into the sea, and our boats picking them up, we succeeded in saving two hundred. All the rest—I do not know how many—were drowned. We finally got these two hundred persons safely on board the *Daniel Webster*.

Here we discovered other difficulties, and it seemed, for a time, as if starvation might do the work that had been denied the waves. There was, also, the question of accommodations; but we solved this problem by taking some of our extra sails and tarpaulin and rigging up a protection for them on deck and in the holds, so that we made them all fairly comfortable. The problem of food was far more difficult. We simply had no food, the captain said. There was hardly more than enough for the crew and passengers of our own vessel, as the delay caused by the rescue and the departure from our course had made an extra demand upon supplies.

Here a happy thought occurred to me. We happened to be carrying a cargo of corn-meal. I had heard that the Irish, in one of their famines, had been fed with corn-meal, learning to eat and even to like it.

"Open the hatches!" I cried, with the enthusiasm of the philosopher who cried "Eureka." The problem of food was soon solved. Two of the barrels were cut in half, making four tubs. From the staves of other barrels we made spoons, and from the meal we made mush which the half-starved men, women, and children ate with great relish. They lived on it until we got them safely landed on English soil, the entire two hundred persons reaching port without the loss of a single soul.

This was my first service at a rescue, and, of course, I was proud of it. Captain Howard received a handsome medal from the Life Saving Society of England, and the incident greatly increased the reputation of our packets.

The *Daniel Webster* continued to sail, Captain G. W. Putnam succeeding Captain Howard, between Boston and Liverpool until 1856 when Enoch Train & Co. failed. After this failure Captain Gaius Sampson took command. When returning from a voyage to Calcutta, he was lost overboard from the quarter deck, and although efforts were made by mate Goodsteep to save him, he sank before help reached him, the ship at the time being off St. Helena Isle.

CHAPTER XV

FEW enterprises appeared more discouraging at the commencement, than the attempt made by Enoch Train of Boston, in organizing a regular line of packets to sail between that port and Liverpool. Twice attempts had been made before, under the auspices of men of great wealth and undoubted enterprise, but owing to various causes, failed. When Mr. Train renewed the attempt, the prospect was far from encouraging. Many of Boston's importers had been so long in the habit of receiving their merchandise by way of New York, that they had become convinced that no packets which could be built there, would be able to compete successfully with those of the Empire City. Mr. Train had to dispel this idea; he had to demonstrate that a Boston line of packets equal, if not superior, to any belonging to New York, could be built in New England, before he could obtain the confidence of Boston importing houses. This required a large outlay of capital, and what was of equal

97

importance, the exercise of much discretion in selecting a suitable mechanic to carry out his views.

It was at the instance of Mr. Train that Donald McKay removed from the Merrimac to East Boston, and thenceforward he built every ship in Train & Company's fleet of packets. The *Washington Irving, Anglo-Saxon, Ocean Monarch, Anglo-American, Parliament,* and *Daniel Webster,* all magnificent vessels, were built by him in succession at East Boston, and by their excellent sailing qualities, and the sound condition in which they brought their cargoes, convinced Boston's mercantile community that they had fairly equalled, if not surpassed, the finest New York packets. The reputation of Train's Line, in a few years was so firmly established that notwithstanding the loss of the *Anglo-Saxon* and *Ocean Monarch,* public confidence never deserted it. The same master mind which organized the line soon filled its broken ranks, and kept it ready at all times to meet the public wants. As trade increased, the several ships added to the line were built in proportion to accommodate the trade; and the clipper packet *Staffordshire,* registering 1817 tons, was the largest and most magnificent of them all.

She was designed to be the swiftest vessel in the Atlantic trade. She had three decks, and was 228 feet long on the keel, and 240 feet over all, from the chock over the bowsprit to the taffrail; had 41 feet extreme breadth of beam, 39 feet width at the planksheer, 29 feet

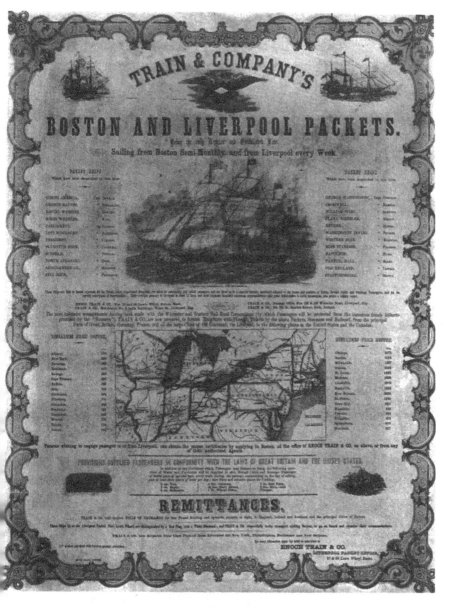

TRAIN & CO.'S POSTER

Showing the *Staffordshire* belonging to this line, also the flag of the firm, list of ships owned, and map showing distances from Boston. This poster is rare. Reproduced through kindness of State Street Trust Co., Boston.

SINKING OF THE "STAFFORDSHIRE"

depth from the upper deck, 20 inches dead rise at half-floor, 3 feet sheer, 1 foot rounding of sides.

Captain Albert H. Brown, who was senior captain of Train's White Diamond Line, was placed in command of the *Staffordshire* and superintended her outfits. He was one of the most capable packet masters sailing out of Boston, one who, by the force of his own character, had raised himself step by step from before the mast to his present position.

The *Staffordshire* was named after the great potteries in England, from which Messrs. Train & Company received so much of their import freight, in recognition of which her stern, where a beautiful arch of gilded carved work spanned its lower division, was ornamented with a manufacturing scene of Staffordshire, and opposite, a representation of Train & Company's Boston storehouse on the end of Lewis Wharf, with a lion's head on each side, and other devices below. Her name and port of hail followed the sweep of this arch and was inside of it.

In the fall of 1851, this handsome clipper packet made a passage from Boston to Liverpool of 14 days, 18 hours, dock to dock.

Shortly afterwards Enoch Train decided to place his clipper on the California route. He engaged Captain Josiah Richardson, who commanded the *Stag Hound* when she sailed from New York and made a record trip to San Francisco of 108 days under strongly adverse conditions.

The *Stag Hound's* performance really established Donald McKay as a designer and builder of clipper ships. It was due however to Captain Josiah Richardson's masterly skill as a navigator, consummate seamanship and rare good judgment that she reached port.

When the *Staffordshire* was being specially rerigged and fitted for this voyage, Captain Richardson was often Donald McKay's guest at his pleasant home in East Boston, continuing a long mutual friendship that was only broken by the former's untimely death.

On May 3rd, 1852, the *Staffordshire* left Boston for California and the East Indies. She made one of the most notable passages of that year, reaching San Francisco, August 13th, in 101 days. She sailed thence for Calcutta. From Calcutta she made the run to Boston in 82 days from the Sand Heads. This we believe is one of the shortest passages on record.

We question whether any ship has ever been more maliciously libelled. When the *Staffordshire* was loading at Boston for 'Frisco, letters were sent to shippers representing that she was unseaworthy, and afterwards a fabricated account of her having been wrecked was published in the Valparaiso papers. Yet she circumnavigated the globe without sustaining the slightest damage, and what is more, beat every vessel which sailed about the same time, not injuring ten dollars worth of cargo.

To test the merits of Astrology and Spiritualism, as mediums of truth, disciples of both were consulted concerning this ship. False reports were made, through prejudice or malice, indicating sickness on board, the death of two or three persons and even referring to the safety of the vessel. Maybe because, for a figurehead, she carried an angelic witch upon the wing robed in white vestments, certain so-called "seers" picked her out for their "Spiritual rappings and tappings." While sailors are a superstitious class, yet they scoffed at such "educated humbug"; but later, developments gave them much for sober reflection, and forecastle judgment eventually placed her in the category of "hoodoo ships."

Certainly the most splendid of Train & Company's ships, beside having proven herself in speed what she unquestionably was in beauty and strength, the *Staffordshire* deserved her sobriquet of the "Queen Clipper Packet of the Atlantic." With her powerful yards painted black, lower masts white, her royal mastheads crowned with gilded balls and spires, ease and gracefulness in the sweep of her lines, she undoubtedly made a beautiful picture as she drew herself out to sea under easy canvas on the 9th of December, 1854, bound from Liverpool to Boston.

It must have been a remarkably fine westerly passage across the Atlantic—only about 20 days to Cape Sable. During a snowstorm, soon after midnight on the 29th—whether scudding or lying to with her head to the northward, is unknown, the *Staffordshire* struck on Blonde Rock near the Seal Islands in the vicinity of Cape Sable, and the wonder is, that under such circumstances any one was saved to tell the tale.

One day before this, Captain Richardson, in coming down from aloft, slipped and fell thirty-five feet, striking on his back, and also badly injuring his ankle, and receiving other serious injury, so he was confined to his cabin when his ship struck several times, and then went off in deep water. Pumps were immediately sounded, and her head was turned around with the intention of beaching her. Twenty minutes after, the wheel ropes parted and she came up into the wind. Then all chances of saving the ship were lost and measures were taken to save the lives. Great difficulty was experienced in launching the ship's boats, and while the last boat was being launched, Captain Richardson's chief mate, Joseph B. Alden, went into the cabin to endeavor to save him. He told him all hopes of saving his vessel were lost. The captain refused Mr. Alden's offer to carry him to the boat, saying the ship was so near shore that she would strike before she would go down. The mate then said, "It is impossible, for she will sink in a few minutes." Captain Richardson then answered, "Then if I am to be lost, God's will be done," which were the last words this brave man spoke!

CHAPTER XVI

"STAR OF EMPIRE" AND "CHARIOT OF FAME," EACH 2050
TONS (OLD MEASUREMENT)
LAUNCHED APRIL-MAY, 1853

"WESTWARD THE STAR OF EMPIRE TAKES ITS WAY"

UPON an arch of gilded carved work, spanning
Star of Empire's stern, was the foregoing line.
For a figurehead she had the Goddess of Fame,
with outspread wings. A trumpet was raised in the right
hand, and her left hand, which was also raised, held a
garland. Her girdle was emblazoned with miniatures of
our distinguished statesmen. The figure was robed in
vestments of white, fringed with gold, its pedestal ornamented with carved floral work.

Of the many ships Donald McKay designed, the *Star
of Empire* and *Chariot of Fame* were the only two built of
the same size and model. In their design and building he
clearly demonstrated that beauty of the highest order
could be imparted to great stowage capacity. In the
outline of their model they had the appearance of first-
class clippers, yet they could carry more deadweight, or

stow more cotton, etc., in proportion to their register, than the fullest-built vessel.

These sister-ships, large three deckers, were designed for Train & Co.'s line of Boston and Liverpool packets, and the following dimensions apply to each: Length on keel 208 feet, and 220 on deck between perpendiculars; breadth of beam, 43 feet, and 27½ feet depth.

With Enoch Train as part owners of the *Star* and *Chariot* were Andrew T. Hall and Benjamin Bangs, both wealthy merchants. Captain Knowles of the latter ship was also financially interested in his vessel.

In keeping with the rapid development of the trans-Atlantic trade, Donald McKay raised the standard of Enoch Train's packet ship service within an incredibly short time, from small, slow-going, clumsy-shaped craft, of about 400 tons, having only the barest accommodations, to large handsome vessels of the clipper type, like *Star of Empire* and *Chariot of Fame*, having splendidly furnished cabins and staterooms and all known conveniences for the comfort of passengers, as well as mechanical equipment and outfits for handling both ship and cargo.

"Sail versus steam" had replaced the natural rivalry of the competitive sailing packet lines. Their frequent quick passages justified claims to supremacy up to this time. Until about 1852 all ocean-going steam craft were side-wheelers and mostly constructed of wood. When one of them, running before a strong, favorable breeze,

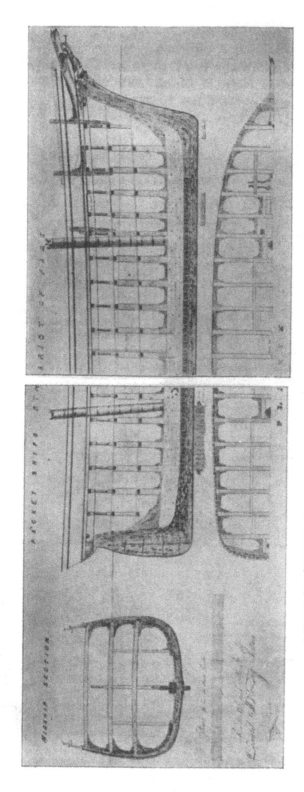

"CHARIOT OF FAME"

Sectional drawings from original in Peabody Museum, Salem, Mass.

"CHARIOT OF FAME"

Sectional drawings from original in Peabody Museum, Salem, Mass.

THIS ILLUSTRATION SHOWS WHAT WONDERFUL ADVANCEMENT DONALD McKAY MADE IN PACKET SHIP CONSTRUCTION FROM 1842 TO 1853

PACKET "COURIER," 380 Tons.
Built in 1842.
Estimated Cost $15,000.

PACKET
"NEW WORLD," 1404 Tons.
Launched, September 9th, 1846.
Estimated Cost $49,000.

CLIPPER PACKET
"STAFFORDSHIRE," 1817 Tons.
Launched, June 17th, 1851.
Recorded Cost $66,000.

CLIPPER PACKETS
"STAR OF EMPIRE," 2050 Tons.
"CHARIOT OF FAME," 2050 Tons.
Recorded Cost $76,000 each.

overhauled in mid-ocean a wallowing side-wheel steamer bound the same way, the joyful shouts and derisive yells of her passengers, as she ranged alongside and swept past, were good for the ears of sailormen to hear. Your old seafarer never possessed a fondness for steamships, as is attested by these two verses, entitled "Roaring Forties" by John Anderson in *Shadows of Sails:*

We've signed in some steamers, some great haughty steamers!
  We've seen all the biggest and best that you boast.
Sink all your steamers! You may call us vain dreamers!
  Poor fools! We had lived ere you came to the coast.

The Forties! The Forties! The wide Roaring Forties!
  With the wind at sou'west and every sheet paid:
We stormed round the world, out and home thro' the forties
  In the noblest creations that man ever made.

Train & Co.'s packets were commanded by a remarkably fine class of men, the best families on the New England coast contributing men to officer these ships. Among the shipmasters in their service, all men of ability and high character, mention should be made of:

Captain Murdoch, who commanded McKay's first Train packet *Joshua Bates*, afterwards the ill-fated *Ocean Monarch* and later *Anglo-American*.

Captain Eben Caldwell, who took out McKay's second Boston and Liverpool packet *Washington Irving*, later commanded the *Plymouth Rock*, and was part owner in both vessels.

Captain Plummer, who, succeeding Caldwell sailed the *Washington Irving* for many years in the Atlantic, as well as California trades, arriving in San Francisco, June 30th, 1851, after a prolonged voyage of 209 days.

Captain Albert H. Brown, who commanded successively the *Anglo-American* and *Parliament*, originally sailed the *Staffordshire;* afterwards, as commodore of Train's White Diamond fleet, took charge of the *Star of Empire.*

Captain Josiah Richardson, who so nobly lost his life while in command of the *Staffordshire* when she sank off Cape Sable.

Captain William H. Howard, of the *Daniel Webster* and *Cathedral.*

Captain Gaius Sampson, commander of the *Parliament, Sunbeam* and *J. Eliot Thayer;* who, after Train's firm failed, took out the *Daniel Webster* on a voyage to Calcutta, and on his return was lost overboard.

Captain Allen H. Knowles, who commanded the *Chariot of Fame* for several years in various trades.

The captains of these packets got credit for the quick runs they made, such as is now only given to our fast steamers on account of their build, model, and the power of their engines, etc. They fully deserved this personal acknowledgement, for their success did not depend, as in the latter case, upon the uniform power of steam, but upon fickle winds and the management of their ships.

As we have drifted, somewhat out of our course, we will let go the anchor and return to our Atlantic packet.

The *Star of Empire* on her maiden voyage from Boston to Liverpool, in May, 1853, crossed in 19 days. Upon a return voyage, August of the same year, though 45 days at sea, out of 830 passengers, only two children died, a less percentage of mortality than might be expected and which showed that her accommodations for passengers were productive of the best results. This ship was equipped with Emerson's ventilators, patent circular air ports in her between-decks and other arrangements to preserve the health of passengers and crew. When voyages were prolonged by adverse winds, such as the *Star* had encountered continually, sailing packets often became floating pest-houses, so her arrival at Boston with so slight a mortality created much favorable comment.

In January, 1854, the *Star* crossed from Boston to Liverpool in 14 days, 15 hours—exceedingly fast time for a packet built primarily for cargo-carrying. This vessel was lost at sea after a too-brief career.

The *Chariot of Fame*, the last ship Donald McKay built for Enoch Train, was said to have been the latter's favorite of all his many packets. She had a reading-room for cabin passengers on her quarter deck, which was considered a great luxury in those days. Her commander, Allen H. Knowles, was not only an able navigator, but a well-educated gentleman held in high esteem by thousands who made the Atlantic crossing with him.

As a White Diamond packet the *Chariot* continued on the Boston-Liverpool run until 1854.

Foremost among the lines striving for supremacy in the Australian trade was the White Star, the same that is plying the Atlantic today. To meet the advantage gained by the Australian Black Ball's possession of the *Sovereign of the Seas* and other McKay flyers, the White Star chartered the American clippers, *Chariot of Fame*, *Blue Jacket* and *Red Jacket*. In this service the *Chariot* carried many hundreds of emigrants to Australia and New Zealand and made some good passages, on one of which she went out to Melbourne from Liverpool in 66 days.

She then engaged in general trade, and, still under Captain Knowles, sailed from New York to San Francisco, arriving there May 13th, 1858, after a passage of 125 days. Her next run over the same route took 117 days. She arrived August 3rd, 1859, but remained at San Francisco some weeks. Shortly afterwards the *Chariot* was chartered for guano, Chincha Islands to Hampton Roads, arriving at her destination in June, 1860. Under command of Captain Hubbard, she loaded at Baltimore for another voyage to California and anchored in San Francisco Bay, December 27th, 1860, after a long stormy passage of 143 days.

The *Chariot of Fame,* like many other American ships, was sold abroad and another flag flew at her masthead. It is said that she came to her end in January, 1876, being abandoned when bound from the Chincha Islands to Cork.

Thus closed the career of the last McKay-built ship of Enoch Train's famous White Diamond fleet. These "Boston Packets," owned and operated by this enterprising New England merchant, comprised principally of fine ocean carriers designed and constructed by Donald McKay, transported thousands of emigrants from poverty and want in Europe to America's Land of Promise, where they gained wealth and enjoyed prosperity. Train's packets not only rendered illustrious service to the commercial and shipping growth of Boston and New England, but proved a largely contributing cause to the successful development of our Western country as well as the Dominion of Canada.

PART III

CALIFORNIA CLIPPER SHIP ERA, 1850–1853

"Time was when noble clipper ships graced the seven seas,
  And the house flags of McKay-built ships were whipped
    by every breeze;
  Where those house flags used to flutter, o'er snow-white
    sails unfurled
  Now the kettle-bellied cargo tubs go reeling around the
    world."

CHAPTER XVII

THE Clipper Ship is an American invention, the initial three-masted clipper ship being the *Rainbow,* a vessel of 750 tons launched from the Smith & Dimon yards, New York, in January, 1845, from plans and designs made more than two years before by John Willis Griffiths, whose genius revolutionized the science of naval and marine architecture, by contributing this—the pioneer clipper—to the needs of mankind.

The Baltimore clippers were built before this time to be sure. They were generally schooner rigged; sometimes rigged as hermaphrodite brigs and brigantines, *but they never carried more than two masts*, so that the Baltimore clipper was not, strictly speaking, a clipper ship.

These Baltimore clippers, however, were fast sailers. They had round pumpkin-like bows, carried high with an excessive overhang. From the bow, the lines ran aft gracefully to a long low-lying stern. The beam of these vessels was all out of proportion to their length, the greatest beam being well forward of the vessel's centre. This excessive beam gave them the ability to stand a heavy press of sail, but they could not work to windward efficiently. Their underbodies were modelled somewhat after a codfish.

And from these prerevolutionary war craft were taken the lines of practically all the merchant ships of this country, previous to the advent of the *Rainbow*.

The statement has often been made that Donald McKay was a disciple or follower of John Willis Griffiths. There is no doubt each of them had been impressed by the *Ann McKim*, a ship of 493 tons register (a large vessel for those days) constructed at Baltimore. She was really an enlarged clipper schooner rigged as a ship and she may have suggested the clipper design in vessels of ship rig; and as she sailed in the China trade and was purchased about 1837, by Howland & Aspinwall (who owned the *Rainbow*) it is not unreasonable to suppose she may have exercised some influence upon the minds of these two young enterprising marine architects, when she was undergoing repairs at one of the shipyards located along the East River, near to where both young men at that time were employed.

As our maritime trade grew, New York and Boston shipbuilders strove to produce fast ships, and so the

American Clipper was evolved. Although Donald McKay did not originate the clipper ship he was the man who made it famous, almost at a bound we may say. His advance production of a vessel of the "extreme" clipper class proved a notable contribution to America's prestige as a maritime nation, and the appearance of his clipper ship *Stag Hound* caused a sensation.

What was learned and evolved by the construction of the California clipper ships practically established on the Atlantic seaboard of the United States a school of world-experts in the art of designing and building ships of wonderfully swift sailing ability. The ingenuity and skill of American shipbuilders combined with the un-rivaled seamanship of fearless Yankee skippers placed Americans in the lead as sailing racers of the sea. Our clippers were built to carry all the spread of sail possible, and their captains became noted for the way they would drive their vessels, never reefing a yard of canvas night or day, except in a heavy gale.

But no story of the California clipper ships would be complete without mention of Lieut. Matthew Fontaine Maury, U.S.N., the celebrated American hydrographer, who revolutionized the methods of sea-borne commerce and well deserves to be known as "The Pathfinder of the Sea."

When our merchant navigators sailed their ships a good deal by "rule of thumb," Lieutenant Maury, the projector of the wind and current charts, did more to arouse them from their lethargy than any other living man, for he made them co-workers with him in recording the temperature of air and water, the strength and direction of the wind, and in noting any natural phenomena so constantly occurring at sea.

He furnished them with blanks to fill up, and with valuable charts, and in a very short time organized a thoroughly interested corps of nautical observers, who by their combined efforts did much to shorten the passage to South America and around Cape Horn by many weeks. The daily newspapers also contributed in the general "Marine Awakening," for they, too, furnished ship masters with blank logs to fill in, and published them when returned; and thus encouraged, and with a laudable desire for notoriety, the shipmaster sought extraneous aid.

In charting the winds and ocean currents Maury gave a new science to the world. He dispelled the myths of navigation and showed shipmasters the way to save time and money. Without Maury's "Sailing Directions" our clippers could never have made such fast voyages.

The effort, three-quarters of a century ago, to construct a fleet of sailing ships for the San Francisco trade continued as a Yankee tradition, that kept America the peer of all nations as a producer of sea-going craft. To Donald McKay, America owes the honor of producing, during the Clipper Ship Era, the best, swiftest, most beautiful and largest merchant ships in the world, which contributed immeasurably towards her national growth and prosperity.

CHAPTER XVIII

"STAG HOUND," 1534 TONS
LAUNCHED DECEMBER 7TH, 1850

WHEN the demand for quick passages around the Horn was at its inception, Donald McKay concluded a contract with George B. Upton and the prominent East India merchants, Sampson & Tappan, both of Boston, to build in sixty days from the signing of the documents, the pioneer of the "extreme" clipper fleet—vessels that were distinguished then and since by the name "California Clippers." Truly "speed was the spirit of the hour!" This, his first notable contribution to the Cape Horn-California fleet, was the *Stag Hound* of 1534 tons register, only carrying her registered tonnage of dead weight. She was designed, modelled and draughted by Donald McKay,—who also draughted her spars and every other scientific detail about her.

Not only was she the largest merchant vessel ever built up to the close of 1850, but her model could be said to be the original of a new idea in naval architecture. Although longer and sharper than any other merchantman

118

in the world, her breadth of beam and depth of hold were designed with special reference to stability. Every element in her had been made subservient to speed; she was therefore Donald McKay's beau ideal of swiftness; for in designing her, he was not interfered with by the owners. He alone remained responsible for her sailing qualities.

This magnificent ship had incited the wonder of all who viewed her on the stocks. A throng estimated at over ten thousand gathered about the McKay shipyard at the foot of Border Street, East Boston, to witness her launch, upon the bitterly cold day in December indicated above, and it was feared she would be unable to slide down the ways on account of the tallow freezing upon them. Nevertheless, when the dog shores were knocked away she moved so rapidly that the shipyard foreman hurriedly seized a bottle of Medford rum he had for the launching and smashed it across her forefoot, excitedly shouting out "*Stag Hound*, your name's *Stag Hound!*" and lost his hat overboard in the excitement.

A carved and gilded stag hound, represented panting in the chase, and carved work around the hawse-holes and on the ends of her cat-heads, comprised her ornamental work about the bow. She had neither head boards, nor trail boards, and could be said to be naked forward, yet this very nakedness, like that of a sculptured Venus, true to nature, constituted the crowning element of her symmetry forward. As she was five feet higher forward than aft, she sat upon the water as if ready for a spring ahead. Broadside on, her great length, the smoothness of her outline, and the buoyancy of her sheer, combined with the regularity of her planking, and the neatness of her mouldings, impressed upon the eye a form as perfect as if it had been cast in a mould. She was planked flush to the planksheer, and its moulding was carried from the extreme of the head round her stern. Her stern was elliptical, finely formed, and very light. The eye directed along her rail from the quarter to the bow, perceived that her outline at the extreme was as perfect as the spring of a steel bow. The planking along the upper part of the run was carried up to the line of the planksheer and there terminated, and this was done, too, without any irregularity in the width. Below, the planking from the opposite sides met, and the butts formed a series of plain angles down to the stern post. Her run was rounded, not concave like that of most ships, and at the load displacement line was apparently the counterpart of the bow, for her greatest breadth of beam was about midships. An idea of the smallness of her stern may be formed from the fact, that at 8 feet from the midships of the taffrail, over all, she was only 24½ feet wide. The stern projected about 7 feet beyond the stern post. A stag, her name and other devices, neatly executed, ornamented her stern.

Her keel was of rock maple and oak, in two depths,

which, combined with the shoe moulded 46 inches, and sided 16. The scarphs of the keel were from 8 to 10 feet in length, and bolted with copper, and the parts of the keel were also bolted together with the same kind of metal. Her top-timbers were of hackmatack, but the rest of her frame and bulwark stanchions were of white oak. The floor timbers on the keel were sided from 10 to 12 inches, and were moulded from 14 to 16, alternately bolted with inch and a quarter copper through the keel, and through the lower keelson and the keel. She had three depths of midship keelsons, which combined, moulded 42 and sided 15 inches. She had sister keelsons 14 inches square, bolted diagonally through the navel timbers into the keel, and horizontally through the lower midship keelson, and each other. Her hold stanchions were 10 inches square, and were kneed to the beams above and to the keelson below, so that their lower arms formed almost a rider along the top of the keelson. Including their depth and the moulding of the floor timbers, she was over 9 feet through "the back bone."

The ceiling on her floor was 4½ inches thick, square bolted, not tacked on with spikes, and all the ceiling from the bilge to the deck in the hold was 7 inches thick, scarphed and square fastened. She had also a stringer of 12 by 15 inches, upon which the ends of the hanging knees rested, and were fayed. The hanging knees in the hold were sided from 10 to 11 inches, were moulded from

2 feet to 26 inches in the throats, and had 16 bolts and 4 spikes in each. In the between-decks the knees had 18 bolts and 4 spikes in them, were sided about 10 inches, and moulded in the angles from 20 to 22 inches. The hold beams averaged about 16 by 17 inches, and those in the between-decks 10 by 16, and were of hard pine. She had a pair of pointers 30 feet long in each, 3 breast-hooks and 3 after-hooks, all of oak and closely bolted. Her hold was caulked and fayed from the limber boards to the deck.

The between-decks were 7 feet high; their waterways were 15 inches square, the strake inside of them 9 by 12 inches, and the two over them combined 10 by 18. Her between-deck stanchions were of oak turned, secured with iron rods, through their centres which set up below. The breast hook in this deck extended well aft, and was closely bolted. Her deck hooks, and the hooks above and below the bowsprit, were very stout and well secured.

The upper deck waterways were 12 inches square, and the two strakes inside of them each 4½ by 6 inches let over the beams below, and cross bolted. The planking of both decks was 3½ inches thick, of white pine.

Her garboards were 7 inches thick, bolted through each other and the keel, and upwards through the timbers and the floor, and riveted. The strakes outside of them were graduated to 4½ inches, the substance of the planking on the bottom, and she had 16 wales, each 5½ by 6 inches. As before stated, she was planked up flush to the covering board. Her bulwark stanchions were 8 by 10 inches, and the planksheer and main rail each 6 by 16 inches. Her bulwarks, including the monkey rail, were 6½ feet high; and between the main and rack rails she had a stout clamp, bolted through the stanchions, and vertically, through both rails. The boarding of her bulwarks was very narrow, neatly tongued and grooved, and fastened with composition. More than usual care had been bestowed in driving her bilge and butt bolts, and the treenails, in order to obtain the nicest possible state of finish outside, combined with strength through all.

She was seasoned with salt, and had ventilators in her decks and along the line of planksheer, fore and aft, and also in the bitts. Her bowsprit and windlass bitts, also the foretopsail sheet bitts, were all of choice white oak, and strongly kneed above and below. Her maintopmast stays led on deck, and set up to the bitts before the foremast.

She had a topgallant forecastle, the height of the main rail, in the after wings of which she had water closets, for the use of the crew.

Abaft the foremast she had a house 42 feet long by 24 wide, and 6 high, which contained spacious accommodations for the crew, and other apartments for a galley, store rooms, etc. The upper part of the house was ornamented with panels, which looked neat.

Her cabins were under a half poop deck, the height of the main rail, and had a descent of three feet below the main deck. Along the sides, and round the stern, the poop was protected by an open rail, supported on turned stanchions. On this deck she was steered, and she had a patent steering apparatus, embracing the latest improvements. The deck itself was 44 feet long, and in its front, amidships, was a small square house, or portico, to the entrance of the cabins.

The after cabin was 32 feet long by 13 wide, and 6 feet 8 inches high. Its after division was fitted into a spacious stateroom with two berths, and was admirably adapted for the accommodation of a family. Before this there was a water closet on each side, then a stateroom, before that a recess of 8 feet on each side, and then two staterooms. The sides of the cabins were splendidly finished with mahogany Gothic panels, enamelled pilasters and cornices, and gilded mouldings. It had a large skylight amidships; and every stateroom had its deck and side light also.

The forward cabin contained the captain's stateroom, overlooking the main deck, on the starboard side; it also contained the pantry, and staterooms for the three mates and steward. It was 12 by 18 feet, and neatly painted and grained; and lighted the same as that abaft.

Inside the ship was painted pearl color, relieved with white, and outside black, from the water's edge to the rail.

She had patent copper pumps, which worked with fly wheels and winches,—a patent windlass, with ends which ungeared, and two beautiful capstans, made of mahogany and locust, inlaid with brass. She had a cylindrical iron water tank of 4500 gallons capacity, the depth of the ship, secured below the upper deck, abaft the mainmast, and resting upon a massive bed constructed over and alongside of the keelson. The groundtackle, boats, and other furniture was of the first quality, and every way worthy of the ship.

Although she was sharp beyond all comparison with other ships, still her floor was carried forward and aft almost to the ends, and presented as true a surface to the water, as ever graced the bottom of any vessel of equal length. That she had a long and buoyant floor is evident from her launch displacement. When launched, she drew 10½ feet forward, and 11 feet 6 inches aft, and this too, including 39 inches depth of keel and shoe, clear of the garboards.

It is interesting to note here, that the early Baltimore clippers had sharp floors and sailed with a drag. Some of them drew only 8 feet forward and 16 feet aft,—the midship section, or broadest part of the hull, being at two-fifths the length from the bow, as in the packets and heavy freighting ships. The forward body was full and the after body lean and tapering under water. What an innovation Donald McKay created in designing his initial clipper ship, can easily be arrived at: 1st, by comparison of the *Stag Hound's* launching draught with these figures; 2nd, by taking into consideration her long, sharp bow, her greatest breadth of beam about midships and light, finely formed, elliptical stern.

When the *Stag Hound* arrived in New York, towed there by Boston's historic tugboat *R. B. Forbes*, arrangements were made for her to soon make her loading berth at the foot of Wall Street. She was commanded by Capt. Josiah Richardson, a gentleman of sterling worth as a man, and a sailor of long-tried experience who had been in the employ of Enoch Train & Co., Boston, in trans-Atlantic traffic. She was immediately made the subject of much comment. Those who had not seen her on the stocks imagined from her sharp appearance on the water that she would bury in heavy weather; but this impression was erroneous, for she was, in fact, very buoyant for her tonnage; and, what is more, she proved to be a remarkably dry vessel on deck in the worst of weather. No ship so heavily sparred had ever been seen in the port of New York. Her sail-carrying power caused wonder, as she spread over 8000 yards of canvas.

Compared with previous marine architectural productions, all her innovations tended to show the extent and degree of the departure that the designer of this pioneer clipper made. It was dubbed by the knowing ones to be rash, stupid, etc., and in fact, so much was said in criticism of the *Stag Hound* that the marine under-writers on her first voyage charged extra premiums to insure her, for she was thought to be an extra hazardous risk. California freights were booming then and the *Stag Hound* quickly filled up at $1.40 per cubic foot. Her first cost was paid before she cleared for San Francisco.

She had quick despatch and left New York, with a crew composed of 36 able seamen, 6 ordinaries and 4 boys, February 1st, 1851. One can well imagine the *Stag Hound* leaving her anchorage off the Battery with a strong westerly breeze and the ebb tide behind her, and as her sails are hoisted and set, she slips down Ship Channel and out by Sandy Hook at a ten knot clip, while her crew are lustily lifting their voices to this most apropos topsail chantey:

> Down the river hauled a Yankee clipper,
>   And it's blow, my bully boys, blow!
> She'd a Yankee mate and a Yankee skipper,
>   And it's blow, my bully boys, blow!
>   Blow ye winds, heigh-ho,
>   For Cal-i-forni-o,
>   For there's plenty of gold,
>   So I've been told,
>   On the banks of the Sacramento.

That old song tells the story of the original purpose of these matchless sailing ships. It is claimed that California caused them to be built. Granted, they were created to meet a special demand for that trade,—the fact remains that the Yankees' (an overwhelming proportion of the Cape Horn Fleet captains were New-Englanders) superiority in managing mountains of towering canvas such as no other skippers would dare spread before a stiff breeze, was a prime factor in the achievement of their unparalleled successes.

When six days out in a heavy southeast gale, the *Stag Hound's* maintopmast and three topgallantmasts came down by the run. She was without a maintopsail for nine days and it was twelve days after the accident that topgallants were ready to set. Although she was forced greatly to the eastward by the terrific gales, her crossing the equator 21 days from Sandy Hook deserves more than passing notice. She arrived at Valparaiso in 66 days, under jury rig on April 8th, 1851, from which port Captain Richardson wrote her owners, Sampson & Tappan, Boston:

Gentlemen:–

Your ship the *Stag Hound* anchored in this port this day after a passage of sixty-six days, the shortest but one ever made here; and if we had not lost maintopmast and all three topgallantmasts February 6, our passage doubtless would have been the shortest ever made. . . . The ship is yet to be built to beat the *Stag Hound*. Nothing that we have fallen in with as yet could hold her play. I am in love with the ship; a better sea boat, or better working ship, or drier, I never sailed in.

"STAG HOUND" HOVE-TO OFF CAPE HORN.

From painting by Charles Robert Patterson.

The *Stag Hound* was at Valparaiso five days; thence had a passage of forty-two days to San Francisco, light winds and calms prevailing. Was twenty-one days from Valparaiso to the line and passed within the Golden Gate twenty-one days afterwards. Her total time from New York was 113 days; total days at sea, 108. A record voyage that did much to establish Donald McKay as a builder of fast ships. Best day's run 358 nautical miles.

She sailed from San Francisco, June 26th; went to Manila and thence to Canton, where she arrived September 26. Loading there a cargo of tea on Owner's account, she sailed October 9; cleared Sunda Straits 31, then had twenty-six days to Cape of Good Hope. She arrived at New York after a passage of ninety-four days.

The results of her first voyage were very satisfactory. Her outward cargo to San Francisco had been secured at unprecedently high rates, so her freight list exceeded $70,000. Her homeward cargo of tea was sold at auction in New York by Haggerty, Draper & Jones, and a few days after, when the earnings of the voyage had been computed it was found that the ship had been absent ten months, twenty-three days and that she had paid for herself and divided among her owners over $80,000. And she proved to be all that her Captain desired, fast, stiff and dry in heavy weather.

At New York Captain Richardson left the ship to take the *Staffordshire* on a voyage to San Francisco and the

East Indies.   Captain C. F. W. Behm took command of the *Stag Hound*.

She commenced her second voyage from New York March 1, 1852, 4 p.m., crossed the line twenty-six days out; passed *St. Roque* twenty-nine days out; thence twenty-nine days to 50 degrees south; twelve days from 50 to 50; and was at the latter point May 9, sixty-nine days out.   On May 4, she was off the Horn, sixty-four days out.   From 50 degrees to the line was twenty-three days, the crossing being on June 1, ninety-one days out; thence to San Francisco nearly thirty-four days.   She arrived July 4; passage, 124 days, during nearly all of which she had had light or baffling winds.   She had three sky-sails set for eighty-three days, she was within 1000 miles of San Francisco for twenty days.

She arrived July 4, but only commenced discharging on the 6th.   Her cargo was all out in nine working days, no steam power being used, very quick dispatch considering labor conditions, etc., then prevailing at San Francisco.   Taking in ballast and water she cleared 20th and sailed 21st, having been in port just thirteen working days.   This was at the time unprecedented. She proceeded to sea under sail, directly from Pacific Street wharf, the papers of the time giving a graphic description of the scene at the wharf.   On July 31, just ten days out, the *Stag Hound* passed close to Honolulu, from whence came the report that with a fair wind, all sail set, she was going off towards China in fine style. She arrived at Whampoa September 6, a trifle under forty-six days passage. On this run the *Stag Hound* beat the clipper ship *Sea Serpent* which sailed with her, two days to Honolulu and nine days to China.

Sailed from Canton September 25, 1852, against the monsoon, in company with the New York clipper ship *Swordfish*. She passed Anjer October 18 and cleared the Straits next day. Thence had thirty-three days to the Cape and arrived in New York December 30, in ninety-five days from China. The passage of the *Swordfish* was ninety days upon which voyage she earned the sobriquet "Diving Bell" and was shunned afterwards by "knowing ones" going to sea. Her gain was made mostly in the run down the China Sea, though she beat the *Stag Hound* one day from Anjer home.

The *Stag Hound* left New York on her third voyage February 25, 1853. She crossed the line in twenty-one days, six hours. She was obliged to put in to Cumberland Bay, Juan Fernandez, for water and was detained off Forland by a gale from May 10 to 14. Left Juan Fernandez May 15; thence twenty-six days to San Francisco, where she anchored July 1, passage 126 days, net 122 days. Had skysails set for eighty-one successive days, having had light winds nearly all passage.

Leaving San Francisco July 16, she arrived at Hong Kong September 13; passage, sixty-one days. On this voyage she experienced a typhoon August 30, 1853— latitude 20 N., longitude 128 E., which continued until September 4th without abating, the ship being under bare poles the whole of the time. The gale was the most violent Captain Behm had experienced for twenty years. The ship sustained no damage.

She sailed from Canton October 24 for New York, arriving January 21, in eighty-nine days,—a quick passage.

Sailing on her next voyage from New York April 27, 1854; she arrived at San Francisco August 14, passage 110 days. Captain Behm was still in command.

Leaving the last named port for Hong Kong, August 25, arrived there October 14, passage forty-nine days. A round trip to Manila followed, from whence she returned to Hong Kong December 19, 1854. Then went to Shanghai, where she loaded for London. Arrived at London August 28, ninety-one days from Java Head.

From London she sailed to Hong Kong. Date of sailing not at hand. Arrived out March 11, 1856. Sailed from Canton April 21 and arrived at New York August 21, 1856; passage 122 days.

Left Boston January 4, 1857; she was now under command of Captain Peterson, and arrived at San Francisco April 22, 1857, in 108 days. Sailed May 15; arrived at Hong Kong July 5, passage fifty days. Sailing from Foochow August 13 she arrived at New York December 4, 113 days.

Sailing from Boston, February 6, 1858, under command of Captain Samuel B. Hussey, she crossed the line in eighteen days, not a record but very good time. From March 24 when she passed through the Straits of Le Maire, to April 20, she experienced stormy weather, it being one continual gale, with high seas, hail, snow and rain, thunder and lightning. Arrived at San Francisco, June 7, after a passage of 121 days. A newspaper statement was that one of her anchors showed a fluke bent up against the stock, hammered up by the force of the seas off Cape Horn.

Another voyage to China followed and she reached Hong Kong, September 17, fifty-eight days out. She laid up there for a time and this was her last appearance in San Francisco.

In due course of time George B. Upton sold his share in the *Stag Hound* to Messrs. Sampson and Tappan, who in turn transferred her to the firm of E. & R. W. Sears, merchants and ship-owners, of State Street, Boston.

This was in 1860. In the spring of the following year she was thoroughly overhauled and placed on a New York berth for a voyage to London, expecting upon arrival at that port to obtain an Australian charter, and in order to facilitate matters a first-class skipper, Captain Lowber, was placed in command. She was ready for sea on March 2.

Now it so happened that James Gordon Bennett,

hearing of the expected sailing of the *Stag Hound*, conceived an idea that there might be a possibility of the vessel reaching London before the regular mail steamer, and so he made arrangements with the Captain and owners to take a copy of Lincoln's first inaugural address, promising that if the *Stag Hound* delivered the message in advance of the mail packet a handsome prize would be given.

The *Stag Hound* sailed from New York early Sunday morning, March 3, with the precious document on board. The mail steamer *America* left Boston harbor on the afternoon tide of March 6. The *Stag Hound* arrived off Gravesend March 18. Captain Lowber hastened forward the telegraphic message to London of the inaugural address, as directed by Mr. Bennett, and was over half-way through when he was informed that they were receiving a similar message via Liverpool from the *America*, which had just arrived, and the *Stag Hound's* message was stopped short. Captain Lowber claimed the reward, as it was proved conclusively that his message was in the hands of the telegraph operator ahead of the one sent by the *America*, and he was awarded the prize.

The *Stag Hound* did not succeed in getting her expected Australian charter, and, as the owners wanted her in the Pacific as soon as possible, she was sent to Newcastle, England, for a load of ballast coal, and then ordered to San Francisco. She was now commanded by

Captain Wilson. All went well until they got off the coast of South America, when the ship took fire from spontaneous combustion and so rapid was the spread of the flames and the density of the smoke that the captain was unable to rescue any of his private property from his cabin.

On beating a retreat from that section he thought of the great American ensign that was stowed away in the flag locker, and, as he passed, he made sure of the Stars and Stripes, and winding it around his body, jumped into the boat. The ship was then many miles from the coast, and, after no little exposure and suffering, the captain and crew arrived safely at Pernambuco; from thence the captain made his way as well and as rapidly as transportation at his command would allow him, to Boston.

On entering the office of the Messrs. Sears, he threw down a large bundle, stating that it contained the flag, and that was all that was left from the *Stag Hound*.

CHAPTER XIX

MODELS—THEIR ORIGIN, USE AND PRACTICABILITY—MODEL
OF THE "STAG HOUND"—WHAT BECAME OF THE
McKAY MODELS

BEFORE engaging the attention of our readers in the subject of the model of Donald McKay's first clipper ship, a short account respecting the origin, use and practicability of these models during the sailing ship period, should prove quite interesting.

In the primary design of a sailing ship the model is obviously the complement of the *draught*. It is to a three-plan mechanical drawing what statuary is to a written description—one the shadow, the other the substance. It takes years of practical training to read a mechanical drawing, while a child can form instantly some conception of a model. The model of a ship shows all proportions—length, breadth and depth—each in its relative position, together with the exact amount and degree of curvature in its rotundity.

It is contended that the old-time ship models were made "by the eye." Even so, it is difficult to find a more correct instrument—a more wonderful power exerted by any sense—than that of the human eye. Our success which staggered all Naval Europe in the War of 1812 was owing wholly to the superiority in the form of the vessels that American mechanics built, and built in a hurry, too. They sailed faster, worked quicker and always outmaneuvered and outfought the ships of England. The superior excellence of our ships was obtained wholly by the use of the waterline model in the designing of them.

Block models of vessels are as old as ships themselves, but the waterline model, which combines all the advantages of the draught with many others, was a Yankee invention and the first waterline model ever made is labelled "The Original Ship Model, made by Orlando B. Merrill, of Belleville, Newburyport, Mass., in 1796."

However much a shipbuilder and designer and the workmen employed upon a vessel were entitled to praise, the owners, after all, had to foot the bills. To their taste for adopting the model, the builder was indebted for an opportunity of showing his skill. Nothing more clearly indicated the taste of a mercantile community than its ships. A merchant selected a model and formed a contract to have it built after, and if the contract was fulfilled, here the builder's responsibility ended. The success or failure of a ship, under such circumstances, ought to have been attributed to the merchant alone. This system of building was common in all our large

Atlantic seaports, so that a shipbuilder rarely had an opportunity to show his skill as a designer. As a general rule therefore, the merchants, not the mechanics, ought to have been responsible for the qualities of their ships. Yet in almost every instance where our shipbuilders had an opportunity of displaying their skill, the result, as in the case of the *Stag Hound*, proved most satisfactory.

The model is not used today as formerly. In fact, the half-model,—showing one side of a vessel from the midship line—is seldom used nowadays as a standard by which to measure the form of a vessel as well as her sea-going qualities, etc.

Whether or not Donald McKay made models of all his ships, we know not. There is no doubt a large number of them were thus reproduced. How he divided certain models, with a saw, when dissolving his partnership with William Currier has already been recited. What became of the rest is now going to be told. When Mr. McKay retired to a farm at Hamilton, Mass., his models were taken from their resting places in the mould loft or wherever they were hung up in his office at the East Boston shipyard and conveyed to his new abode. These veritable treasures, maybe for lack of space or appreciativeness, were stored in the barn. Here they met with an ignominious fate! It was not until after Donald McKay was dead and buried that his eldest son, Cornelius W.— also a designer and builder of ships—discovered that

Clipper Ship "STAGHOUND" Designed and Built BY DONALD McKAY 1850.

MODEL OF "STAG HOUND" IN POSSESSION OF THE MARINE MUSEUM, BOSTON.

Extract from Letter of Cornelius W. McKay, when Presenting "Stag Hound" Model to Captain Arthur H. Clark.

"I take pleasure in presenting to you for acceptance the model and drawing of the American Clipper Ship *Stag Hound*. As a lover of and expert in all things marine and nautical, you will, I think, estimate this data of an art almost forgotten at its real worth. The model is made to the ship's 'moulded' dimensions, on a scale of 4 feet to the inch. The water lines are 2½ feet apart. Of course, to the rotundity, or shape of this model the outside planking should be added, also the keel. The bottom planking was 4 inches and the wales 5 inches thick. The keel was built in two depths and was 30 inches deep. Some few 'fancy' models of my father's ships may be laying round somewhere, but the "working" models were all destroyed a few years ago. I know of no other *correct* model of a clipper ship built by my father in existence than this one of the *Stag Hound*. I made it myself from the lines of the ship as they were laid down on the moulding loft floor. The first, or the 'working' model, of a ship was most always altered more or less in the process of laying down in the moulding loft, and builders kept records of their productions in the shape of tables of offsets taken from the mould loft floor and recorded in books. They seldom made a correct model after the lines were faired and 'proved' for the making of the moulds; and owing to the jealousy existing in those times, builders never gave a correct model away."

Reproduced through courtesy of the Boston Marine Museum.

those wonderful conceptions depicting the *Flying Cloud*, *Sovereign of the Seas*, *Westward Ho*, *Romance of the Seas*, and a host more of the famous McKay flyers had been chopped up for firewood!

Only one model, that of the *Stag Hound*, illustrated here, was found at Hamilton and this Cornelius W. McKay forwarded to New York where he lived. Many years afterward, in conversation with the late Capt. Arthur H. Clark, he told the latter how this model then reposed in a sail loft on South Street. Capt. Clark evinced a strong desire to obtain it and they both immediately repaired to the loft where, buried beneath old pieces of sailcloth and the usual sailmaker's débris, they discovered the *Stag Hound* model, which is now in possession of the Boston Marine Society.

In 1869, when Donald McKay's daughter, Albenia, who had married Johannes G. Bodemer, of Zschopau, Saxony, was visiting at East Boston, she prevailed upon her father to give them the model of the *Great Republic* and this is we believe still in her possession.

Only one other escaped being ignorantly destroyed. When the *Glory of the Seas* was completed, Mr. George C. Young, who made her spars, secured possession of the model of this, the last of Donald McKay's clippers. He left it with Albert Low, who had rigged the *Glory* and, in the course of human events, Mr. Young died and the rigging loft of Albert Low came into possession of a Mr.

Connelly, who knowing about the model had it properly taken care of. At Mr. Connelly's death, some while ago, his son George W. took charge of this model and it was from the latter, at his store 14–16 Atlantic Ave., Boston, that the author was able to gain possession of the third and last McKay model known to now exist.

The Peerless *Flying Cloud*, 1782 Tons, 1851. Donald McKay's famous clipper that *twice* sailed from New York to San Francisco in eighty-nine days. From painting by Charles Robert Patterson. Courtesy of R. P. Stevens, Esq.

CHAPTER XX

DONALD McKAY'S bonny clipper has long since
gone the way of all ships, but unlike most
vessels whose brief story passes into oblivion
the *Flying Cloud* has left a bit of maritime history behind
her which is still the talk of men who like to recall events
in the era when clipper ships were the chief means of
commerce upon the Seven Seas.

Happy was the fancy that first led to so dashing a
nomenclature as *Flying Cloud*. Credit for the idea of
creating it is claimed by George Francis Train, junior
partner of Enoch Train & Co., and the following is
extracted from his *Biography:*

When the gold fever was getting the country frantic, and
everyone apparently wanted to go to California, I said to
McKay, "I want a big ship, one that will be larger than the
Ocean Monarch." McKay replied, "Two hundred tons bigger?"
"No," said I, "I want a ship of 2,000 tons." McKay was one
of those men who merely ask what is needed. He said he would
build the sort of ship I wanted. "I shall call her the *Flying
Cloud*," I said. This is the history of that famous ship destined
to make a new era in shipbuilding all over the world.

Longfellow sent me a copy of his poem—"The Building of the Ship," which he had written to commemorate the construction. [See Note relative to this poem being theme of *Great Republic*, etc.] Not only shipbuilders but the whole world was talking of the *Flying Cloud*. Her appearance in the world of commerce was a great historic event. No sooner was the *Flying Cloud* built than many shipowners wanted to buy her; among others the house of Grinnell, Minturn & Co. of the Swallow-Tail Line of Liverpool asked what we would take for her. I replied, that I wanted $90,000, which meant a handsome profit. The answer came back immediately—"We will take her." We sent the vessel to New York under Captain Creesy, while I went on by railway. There I closed the sale, and the proudest moment of my life, up to that time, was when I received a check from Moses H. Grinnell, the New York head of the house, for $90,000.

While on the stocks the *Flying Cloud* was sold to Messrs. Grinnell, Minturn & Co. Enoch Train, the good friend of Donald McKay, often said that there were few things in his life that he regretted more than parting with this ship.

On April 25th, 1851, Duncan McLean, that careful recorder of data, which otherwise might have been forever lost, furnished the following description of this famous clipper ship to the Boston *Atlas:*

If great length, sharpness of ends, with proportionate breadth and depth, conduce to speed, the *Flying Cloud* must be uncommonly swift, for in all these she is great. Her length on the keel is 208 feet, on deck 225, and over all, from the knight heads to the taffrail, 235—extreme breadth of beam 41 feet,

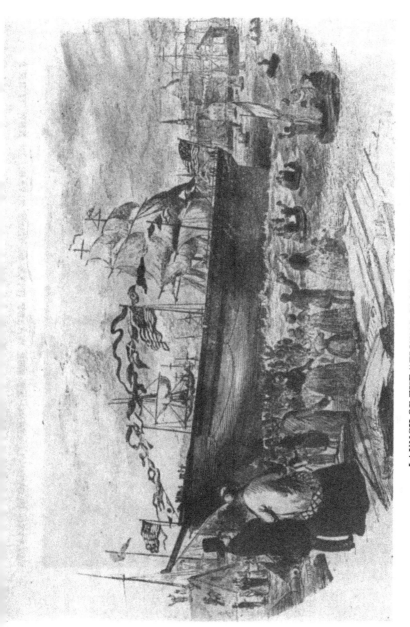

LAUNCH OF THE "FLYING CLOUD" APRIL 15TH, 1851
Reproduced through courtesy of State Street Trust Co., Boston.

"FLYING CLOUD" LOADING AT HER WHARF (PIER 20, EAST RIVER), IN NEW YORK

depth of hold 21½, including 7 feet 8 inches height of between-decks, dead-rise at half floor 20 inches, rounding of sides 6 inches, and sheer about 3 feet.

Apropos of the launching of the *Flying Cloud*, the scene was not unlike one described so beautifully by Henry W. Longfellow in his matchless poem. He was a frequent visitor at the McKay yard and took an active interest in the shipbuilding operations there, as they laid the keel of one beautiful clipper after another and consigned them afterwards to their watery element.

> Then the master,
> With a gesture of command,
> Waved his hand;
> And at the word,
> Loud and sudden there was heard,
> All around them and below,
> The sound of hammers, blow on blow,
> Knocking away the shores and spurs.
> And see! she stirs!
> She starts,—she moves,—she seems to feel
> The thrill of life along her keel,
> And, spurning with her foot the ground,
> With one exulting, joyous bound,
> She leaps into the ocean's arms.
>
> And lo! from the assembled crowd
> There rose a shout, prolonged and loud,
> That to the ocean seemed to say,—
> "Take her, O bridegroom, old and gray,
> Take her to thy protecting arms,
> With all her youth and all her charms!"

However, as we have somewhat varied our course, it becomes incumbent upon us to return to the *Flying Cloud*, that beautiful creation of Donald McKay which, so soon after these events transpired, dashed through the Heads of San Francisco eighty-nine days and some hours out from New York!

During the interval that had transpired since the *Stag Hound* made her much-talked-of appearance, New Yorkers, more especially the habitués of South Street who prided themselves upon their wisdom concerning all things nautical, had become somewhat used to those wonderful specimens of marine architecture that brought fame and fortune to America; in fact, elevated our world-wide commerce to heights since unattainable by American shipping. Consequently the advent of the *Flying Cloud* did not create a sensation. The long, low exquisite hulls of the *Surprise*, *Witchcraft*, and *N. B. Palmer*, among others, had kissed the willing tide in New York harbor. The era of sharp ships was now fairly inaugurated, and with it began a rivalry between Boston and New York shipping interests, which really did much to aid this country's maritime and commercial prosperity. The Boston admirers of this ship were sanguine that she would outsail any vessel in the world, and she fulfilled all their expectations!

It is claimed that Messrs. Train & Co., had appointed Captain Josiah P. Creesy, who hailed from Marblehead,

Mass., to command the *Flying Cloud*. Be that as it may, he was well known in New York, as he had commanded the ship *Oneida* for a number of years in the China and East India trade. He was said to have made five voyages from New York to Anjer between 1845 and 1850, in 90 days or less;—and as he bore a high reputation among ship-owners and underwriters, Messrs. Grinnell, Minturn & Co., tendered him command of their vessel. Future events certainly proved the wisdom of their selection, and that Captain Creesy had cause to be proud in being given command; and Mrs. Creesy, too. She went with him and could most efficiently navigate a ship.

Commencing one of the most eventful voyages ever recorded in the annals of the sea, the *Flying Cloud* swept past Sandy Hook on June 3rd, 1851, bound for 'Frisco! We can imagine what a beautiful sight she presented, running under three skysails, royals, topgallant, topmast and square lower studding sails before a westerly breeze with blue water boiling white along her lee side. This breeze soon freshened to a gale, we are told, as with a bone in her teeth she kept racing away,—but the canvas stayed on her! When only three days out of New York the *Flying Cloud* was partly dismasted by a heavy gale, but while spars were being made and refitted, and new sails bent, the ship went flying on. As excerpts from Captain Creesy's log may interest our readers, we give them below:

June 6.—Lost maintopsail-yard and main and mizzen topgallant masts. June 7.—Sent up topgallant masts and yards. June 8.—Sent up maintopsail-yard, and set all possible sails. June 14.—Discovered mainmast badly sprung about a foot from the hounds, and fished it.

From this time she encountered moderate winds and calms: in four consecutive days she made only 101, 82, 52 and 53 miles, yet nevertheless she crossed the equator twenty-one days out.

Soon after crossing the line a mutiny broke out among the crew, and Captain Creesy ordered several of the men in irons.

"July 11.—Very severe thunder and lightning—double reefed topsails—latter part blowing hard gale, close reefed topsails, split fore and main topmast staysails. At 1 P.M., discovered mainmast had sprung. Sent down royal and topgallant yards and booms off lower and topsail-yards to relieve the mast —heavy sea running, and shipping large quantities of water over lee rail. July 12.—Heavy southwest gales and sea, distance 40 miles. July 13.—Let men out of irons in consequence of wanting their services. At 6 P.M. carried away maintopsail tye and truss band round mainmast, single reefed topsails. July 19.—Crossed latitude 50° south. July 20.— At 4 A.M., close reefed topsails and furled courses, hard gale with thick weather and snow. July 23.—Passed through the Strait of Le Maire. At 8 A.M., Cape Horn north 5 miles distant, the whole coast covered with snow.

She crossed latitude 50° south in the Pacific on July 26, seven days from the same latitude in the Atlantic—a remarkable record.

"July 31.—Fresh breezes and fine weather. All sail set. At 2. P.M., wind south-east. At 6 squally, in lower and topgallant studding sails. 7 P.M., in royals. 2 A.M., in foretopmast studding sail. Latter part strong gales and high sea running, ship very wet fore and aft. Distance run this day, 374 miles. During the squalls 18 knots of line was not sufficient to measure the rate of speed. Topgallantsails set. August 1.—Strong gales and squally. At 6 P.M., in topgallantsails, double reefed fore and mizzen topsails, heavy sea running. At 4 A.M., made sail again. Distance 334 miles. August 3.—Suspended first officer from duty, in consequence of his arrogating to himself the privilege of cutting up rigging contrary to my orders, and long continued neglect of duty. August 25.—Spoke barque *Amelia Pacquet* 180 days out from London bound for San Francisco. August 29.—Lost foretopgallant mast. August 30.—Sent up foretopgallant mast.

And at 11:30 A.M., the *Flying Cloud* dashed through the Golden Gate, 89 days 21 hours from Sandy Hook!"

With sailorlike briefness, entries appear in the log; of sprung masts, split sails, lost spars, and splicing, fishing, and rerigging, to keep things standing somehow; but always the wonderful figures that told the day's run had not been seriously interfered with.

Then the heart-breaking doldrums, when the ship lay for days slatting idly on a breathless sea, trying to fan herself across the line. An entry or so about men in irons and the mate suspended from duty tells us that Captain Creesy had more than the elements to contend with. Little space is given to the struggle with Cape Horn; and soon the ship is flying northward in the Pacific like a mad thing, with solid seas slow to part roaring over the cat-heads while the spume soars to the lower top-sails! A record is made, July 31st: "374 miles run this day."

Calculated, that means an average of fifteen and six-seventeenths knots an hour for twenty-four hours. Not until twenty-three years afterwards was an ocean-going steamship to attain a fifteen-knot speed. Furthermore, her day's run of 374 nautical miles (or knots) is equal to 433 statute miles. (Note: Wherever the terms of mile and knot are employed herein they are understood as meaning the sea measure or nautical mile of 6080 feet, not the statute or land mile of 5280.)

Drive, drive, drive is the order of the day. At last on the ragged horizon the rugged Farallones cut the sky and the *Flying Cloud* rushes through the Golden Gate, famous forever for having made the passage from New York around Cape Horn to San Francisco, in eighty-nine days!

Let the reader change his outlook, by transporting himself to California in those stirring days and there muse awhile. This grand ocean exploit was celebrated in San Francisco with rejoicing! Every American there felt, now that the hazardous voyage around Cape Horn had been made in less than three months, that he was nearer to his old home in the East. Along the Atlantic sea-board, the news was received with enthusiasm, and it was regarded by the press not only as a personal victory for the owners, builder and captain of the *Flying Cloud*, but as a triumph of the United States upon the sea!

No masters of men deserved more credit than the commanders of the Cape Horn fleet. The crews of American ships sailing from our two Eastern ports, which controlled the California trade, were largely composed of the rowdy and knavish class. Their baggage was on their backs, and their purse in every man's pocket. These vagabonds stepped—or were carried drunk or drugged, "shanghaied" oftentimes,—on board an outgoing ship. Then, hey for California! To maintain discipline and sail their vessels, American shipmasters were often compelled to adopt harsh and cruel measures uncountenanced today. Most of the trouble aboard the clippers was due to the tough element that shipped in them as a means of getting to the goldfields of California. The manly decent sailor who knew his work and did his duty was seldom in trouble and always highly prized by intelligent and discerning men like Captain Creesy.

"Off for the mines" went most of Creesy's crew, so the *Flying Cloud* sailed across to China short-handed. The day after leaving San Francisco, with a fresh whole-sail breeze and smooth sea, she is credited with having again sailed 374 miles in twenty-four hours.

Of all the witty things said or written by Mark Twain, no phrase has been quoted oftener than his reply to an alarmist report, "Rumor of my death greatly exaggerated." Little did the author of this witticism know that many years previously another famous American had a somewhat similar experience, although under different conditions. Captain Creesy, during the passage from Canton to New York perused his own obituary in mid-ocean. When his vessel was well across the Indian Ocean, she fell in with a ship outward bound, and in exchange for chickens, fruits and vegetables from Anjer received newspapers from New York, one of which contained the startling announcement of his death on the second day out from San Francisco. This premature publication of his obituary certainly proved advantageous to him, for it stopped an action for damages which his late mate—who had been suspended from duty for cutting up rigging and long neglect of duty—aided by a "sea lawyer," hoped to have ready for him upon his return to New York. The *Cloud* arrived in New York from Canton on April 10, 1852, having made the passage in 94 days.

When California clippers came home from Far East ports or San Francisco at this time, most of them needed a pretty thorough overhauling. It is said that the masts, spars, and rigging of the *Flying Cloud* were fine examples of the skill of her sailors in clapping on fishings, lashings, stoppers, etc. New York and Boston shipbuilders were constantly on the alert to remedy these conditions, however, and every deficiency in construction, rigging or anything else was so carefully noted that continual improvements were made. There was keen rivalry, too, among these master craftsmen, consequently the arts of shipbuilding, sailmaking and rigging, as well as seamanship, reached a state of excellence never previously attained.

Flying Cloud's triumph caused great excitement throughout the country. She had beaten every ship of her time. It is said that her owners, Messrs. Grinnell, Minturn & Co., had her log printed in gold, upon white silk, for distribution among their friends.

On her second voyage from New York she arrived at San Francisco, September 6, 1852, taking 30 days to reach the equator. There had been much talk of the respective merits of the *N. B. Palmer*, a vessel of 1490 tons, built by Jacob A. Westervelt, of New York, and the *Cloud*, especially in view of the fact that the *Palmer* had beaten the McKay flyer about ten days both on the 'Frisco run to China and when returning to New York from Canton. These rivals met off the coast of Brazil. Capt. Charles P. Low in his interesting biography recites:

I had come up with her (*Flying Cloud*), beating her ten days thus far and only forty days out I felt very proud of it. . . . Captain Creesy hailed me and wanted to know when I left New York. I replied, "Ten days after you." He was so mad, he would have nothing more to say. My ship was now at a standstill, and he was going ahead at full speed, and he ran ahead of me. Shortly after I filled away.

At daylight next day, the *N. B. Palmer* was about twelve miles astern and by four o'clock that afternoon, she was no longer in sight.

Both vessels encountered heavy westerly weather off the Horn, *Flying Cloud* arriving at San Francisco, 113 days after leaving New York.

The commander of the *N. B. Palmer* had a hard time of it. His mate had been shot by one of the crew, the second and third mates were of little account, so alone he had to contend with a refractory crew for eighteen days, during which he did not sleep below, but tumbled down in the corner of the house on deck in his wet clothes and got only a few hours sleep during the twenty-four. He managed to put into Valparaiso to get needed supplies and send the mutineers home to be tried for attempted murder on the high seas. Some twenty men deserted the ship, and shipping others cost him further delay, so that the *Palmer* was about three weeks behind the *Flying Cloud* in reaching San Francisco.

From that port she sailed to Canton. Her homeward trip from China, from whence she sailed December 1, took 96 days.

At 4:30 P.M., April 28, 1853, the *Flying Cloud* cast off steamer in New York Bay and made sail. Crossed the

Equator May 15, 17 days from Sandy Hook. To that date she never had skysails in for more than three hours.

For 13 days she had to contend against heavy head gales of wind off Cape Horn. One evening her first officer, Mr. Gibb, reported the foretopmast staysail split to pieces by a sea striking it. Captain Creesy told him to have it hauled down and let it lie, not deeming it safe to send men on the bowsprit, or even on to the forecastle. Later he came again to the cabin, and reported the ship to have come up to N W; exchanged some pleasant words and returned to the deck. In less than five minutes the alarm was given that two men were overboard; captain ordered all hands called, but the night being dark, in a heavy gale of wind, with squalls, the ship going 10 knots, it was impossible to save them. It seems Mr. Gibb went forward and ordered the men to take care of the staysail; went first himself, followed by third mate and four men, and in less than a minute after she put her bows under very deep, taking the first officer and one seaman overboard.

On this voyage the *Flying Cloud* raced against the New York clipper *Hornet*, Captain Knapp, sailing two days before. She anchored off North Beach about 45 minutes ahead of the world's champion, as McKay's matchless clipper was called. The latter had taken 105 days, on this, her third voyage to California and beat the *Hornet* by a day and some hours.

Instead of going to China, Captain Creesy returned to New York. It was on this passage that he reported having twisted the rudder head so badly, in a heavy squall, as to render it useless, and he successfully made port with a temporary apparatus.

Nearly three years had elapsed since the passage from New York to San Francisco had been made in 89 days 21 hours. Though every kind of vessel, ranging from 1000 to 2400 tons, sailed over the same route, that passage had not yet been equalled. The nearest to it was made by *Flying Fish*, another crack McKay ship. It was reserved for the *Flying Cloud* alone to excel her first passage.

"At 12 o'clock, noon, January 22, 1854, this good ship got under way in tow of steamer *Achilles* off foot of Maiden Lane at New York, and proceeded down the bay, bound to San Francisco. Discharged steamer and pilot, made sail and put to sea, and at 6 P.M., passed lightship, wind fresh with cloudy weather," Captain Creesy records in his log.

In chronological order is carefully recorded, strong gales and violent squalls, passing clouds and rain, light airs and fair weather, gentle breezes and hazy weather, fine at intervals, etc., etc.

Arriving at the last entry, we read:

(April) 20th Lat. 37° 18': Long. 123° 54'. Light breeze, hazy weather, at 1 P.M. made Farallones Islands; at 6, took a pilot, and anchored in San Francisco, *after a passage of 89 days, 8 hours!*

The abstract log is as follows:

Sandy Hook to the equator....................17 days
Equator to 50° South.........................25 days
From 50° South in the Atlantic to 50° South in the
    Pacific....................................12 days
To the Equator..............................20 days
To San Francisco............................15 days
                                                     ————
       Total........................89 days

Such an extraordinary passage, like many other great feats, was attributed by some to luck; but the luck, in our opinion, was combined in the excellence of the ship and the skill of her commander.

The *Flying Cloud* sailed after the *Archer*, also an exceedingly fast ship, and led her into San Francisco. In his "Physical Geography of the Sea," Maury describes this McKay clipper's side-by-side sailing qualities so interestingly that we cannot resist quoting it here:

Let a ship sail from New York to California, and the next week let a faster one follow after: they will cross each other's path many times, and are almost sure to see each other by the way. Thus a case in point happens to be before me. It is the case of the *Archer* and the *Flying Cloud* on their last voyage to California. They are both fine clipper ships, ably commanded. But it was not until the ninth day after the *Archer* had sailed from New York that the *Flying Cloud* put to sea, California bound also. She was running against time, and so was the *Archer*, but without reference to each other. The *Archer* with "Wind and Current Charts" in hand, went blazing her way across the calms of Cancer, and along the new route, down through the northeast trades to the equator; the *Cloud* followed after, crossing the equator upon the trail of Thomas of the *Archer*. Off Cape Horn she came up with him, spoke him, handed him the latest New York dates, and invited him to dine on board the *Cloud*, which invitation, says he of the *Archer* "I was reluctantly compelled to decline."

The *Flying Cloud* finally ranged ahead, bade her adieus, and disappeared among the clouds that lowered upon the weather horizon, being destined to reach her port a week or more in advance of her Cape Horn consort.

By the merchants and others in San Francisco Captain Creesy was greatly feted, as this second record-breaking voyage of his clipper roused great enthusiasm.

Even at this late date it may not be amiss to settle the relative merits of *Flying Cloud* and *Andrew Jackson*, which in 1860 sailed from New York to San Francisco in "89 days, 4 hours," eclipsing the *Cloud's* record passage of 89 days, 8 hours. From most reliable authority we can state that the *Jackson* hove up her anchor at New York, December 25, 1859, 6 A.M., discharging her pilot at noon. She received her San Francisco pilot at 8 A.M., March 24, 1860, and anchored in San Francisco Bay 6 P.M. Thus her passage is 90 days, 12 hours, anchor to anchor; 89 days 20 hours pilot to pilot. It must be remembered that *Andrew Jackson's* two best voyages were made in 90 and 100 days, whereas *Flying Cloud* ran the distance twice in 89 days and some hours, and she deserves to stand in first place.

In 8 days, 8 hours from time of arrival at 'Frisco, McKay's white-winged racer was under way, in tow of two steamers bound for Hong Kong. Finding from the strength of the wind and roughness of the sea, together with strong, eddy current that the steamers could do nothing with the ship, and being in a dangerous position, she was obliged to come to with both anchors in 12 fathoms of water, about one mile from shore. The gale continued to blow heavy, and next day the U. S. Steamer *Active* came down to render assistance, owing to a report in circulation at San Francisco that the ship was on shore. Captain Creesy was obliged to decline assistance, as the wind commenced moderating and veering to Northward. Next morning his ship dropped out with the tide, he discharged pilot outside the bar, and with a light northwesterly breeze, freshening all the time he commenced another quick run—this time across the Pacific to China.

At daylight on May 7, 1854, the Captain and crew of the *Cloud* saw the Lema Islands W N W. 35 miles. The pilot was taken aboard at 11 A.M., and the ship anchored in Hong Kong harbor, June 7, at 6 P.M., civil time, 36 days from San Francisco. It had been a fast voyage.

It took some time for the *Cloud* to procure her cargo,

etc., in China for New York. This eventful voyage is so well-described by Walter R. Jones, President of the Board of Underwriters, at New York, February 3, 1855, when he presented Captain Creesy with a silver service set, that we reproduce it in full.

Sir:—On your last passage from China when in command of the celebrated ship *Flying Cloud*, with a rich and costly cargo of delicate goods, the total value of which, probably, amounted to a sum of dollars, you encountered adverse currents, and stormy and foggy weather, which carried your ship upon a coral reef, on the 7th of August last, in the China sea, striking with such severity that her bow was raised out of the water three or four feet, her shoe taken off her keel, and keel itself cut through to the bottom planking causing her to leak badly and to make a great quantity of water.

With a skill that none but a first rate shipmaster possesses, you soon extricated her from her perilous situation, without cutting away her masts or making any other great sacrifice, which is often done, nominally for the benefit of whom it may concern proving very frequently however, to the great detriment of all concerned.

In a very short time you had her afloat ready to proceed, when the important question arose in your mind where you should go; on the settling of which much then depended.

Again your good judgment manifested itself. The expensive and costly ports in the straits were near at hand, you determined to avoid them and no one can say how much you saved to those interested in your valuable ship and cargo, but it is reasonable to suppose that those concerned have been saved at least, thirty thousand dollars and probably much more; in fact no one can possibly tell the extent of saving with much accuracy; all know it has been very large.

At that time your qualifications as a skilful commander again became manifest and you seem also to have combined in yourself the talents of the merchant as well as the ship-master.

After relieving your ship your attention was directed to the next best movement, and in that you rendered us an important service; instead of running your ship into an expensive port before referred to where the positive and known charges would have amounted to a very large sum, you examined the condition of the vessel and the means at your command and although your crew was weak and insufficient you made up your mind to proceed homeward, and with a leaky ship you left the China seas and in a very short time thereafter, to the great relief of the Underwriters you reached this port in safety and with scarcely a damaged package on which a claim could be made upon the Underwriters.

Taking into view the important services you have rendered to the Marine Insurance Companies of the city, by your energetic, prompt, skilful and successful conduct, they have caused a choice and weighty service of plate to be prepared, which I have now the honor, in their name, to present you, as a testimonial of their appreciation of your good conduct, so opportunely and satisfactorily rendered on the voyage referred to, and that you may long and successfully live to enjoy it, is, I can assure you, the ardent wish of all the donors.

We also desire to record our testimonial in your favor, and to make known your example, that the timid may be encouraged and the energetic, sustained and strengthened in a similar course of conduct. In avoiding an entry at a port in the Chinese Seas, and the necessity of discharging and reloading your cargo, you have saved the property from charges to a very large amount, your ship from a long detention, and your crew from the hazards of entering a sickly port, all which it was most desirable for you to avoid, and in doing so you are entitled to our acknowledgments.

Captain Creesy's letter, in acknowledgment of the gift and responding to the "presentation remarks" indicate real appreciation, coupled with an honest modesty respecting his own meritorious services, befitting one of this country's greatest shipmasters. It reads, as follows:

Sir:—I have received your favor of the 3d inst., together with a copy of your remarks at the recent presentation to me of a service of plate. Throughout the voyage of which you speak in flattering terms, I merely did my duty as a Shipmaster, according to the best of my knowledge and ability.

Though for this I claim no praise, I am not insensible to the good opinion of the Honorable Board which you represent, and I am very far, I trust, from being ungrateful for the beautiful and valuable testimonial with which they have seen fit to honor me.

The Sailor, amid the difficulties, dangers and responsibilities of his profession, often feels the need of appreciation and sympathy. These are his best reward and highest encouragement.

From the bottom of my heart I thank you, Sir, and the gentlemen of the Board of Underwriters, for your kind words and rich gift, I shall cherish them while I live, and shall be proud to leave such a legacy to my family.

With great regard, your ob't servant,

(*Signed*) JOSIAH P. CREESY.

The *Flying Cloud* made one more passage to San Francisco. This was in 1855 and she took 108 days. At the end of this voyage Captain Creesy gave up the sea retiring to his home in Salem.

Though far smaller in size than so many of the McKay

clippers, the *Flying Cloud* was destined to overshadow her larger sisters in some respects, even after the advent of the *Sovereign of the Seas*, *Flying Fish*, and *Westward Ho*. Her wonderful record of twice going from New York to San Francisco under sail in 89 days, has never been eclipsed! Built with a view of speed, her sailing qualities were remarkable. And it is meet to state right here, that this famous craft performed wonders only when driven under the masterly skill of that sailor par excellence Captain Josiah P. Creesy, who was, we must also recall, ably assisted by his faithful helpmate, Mrs. Creesy. She constantly proved herself to be a competent navigator, was an excellent nurse and added a real homelike atmosphere to her husband's ship.

The epoch-making era of the American sailing ship was over, instead of continuing their success in the history of maritime commerce, Americans, North and South, made other history—waging an internecine war that eventually drove our flag off the high seas, commercially at least. Financial depressions, aided by changed conditions in the shipping business at this time, had forced many American ship-owners to sell their vessels to foreigners. Conditions were slowly righting themselves, when the firing on Sumter was heard throughout the land, Civil War was on, and the deathknell of American shipping was sounded!

Now we come to the last portion of our gallant ship's

career. About 1863 the *Flying Cloud* was purchased by James Baines of Liverpool, who bought her at a very low price on account of the *Alabama's* depredations, and thus she went under the British flag. He employed her in his Black Ball Line running to Brisbane. In the early seventies she was sold to Smith Edwards one of the founders of Smith's Dock Co., who, it was said, had a great "weakness" for American clippers and who was one of the most enterprising British ship-owners of his day. He did not keep her long, for in June, 1874, she stranded on the coast of New Brunswick when attempting to make St. Johns. She came off without very much difficulty, but she was obviously strained, and it was decided to put her on the patent slip for repairs. While these were in progress she caught fire and was so badly damaged that she was only fit for the scrap heap.

At the outbreak of the Civil War, Captain Creesy was appointed a commander in the United States Navy and assigned to the clipper ship *Ino*. Although registering only 895 tons (a little more than half *Flying Cloud's* tonnage) she carried a crew of 80 men from Marblehead, and on her second cruise, under his command, made the record run of 12 days from New York to Cadiz. He continued to faithfully serve his country during the war. Subsequently commanded the clipper ship *Archer*,—the same vessel he had outdistanced in *Flying Cloud* on that memorable record-breaking passage to San Francisco,—

and he made two voyages to China in her. He died at Salem in 1871. The history of Captain Josiah Perkins Creesy and the *Flying Cloud* is an epic tale of the sea which must live in the annals of our history, so long as brilliant exploits hold a place in the memory of man.

CHAPTER XXI

"FLYING FISH," 1505 TONS
LAUNCHED SEPTEMBER, 1851

THE "*Flying Fish*" and her commander, Captain
Edward Nickels, were among the most popular
that ever sailed from Boston or New York in
clipper ship days. She was owned by Messrs. Sampson
& Tappan, who, with George B. Upton, were the most
prominent Boston ship-owners of their day. The best in
design, construction and equipment was required by them
and they always gave Donald McKay, with a liberality
seldom found in commercial affairs, the utmost freedom
with respect to everything that entered into the building
of their ships, which were also fitted out upon a most
generous scale, with spare gear, stores and provisions.
Among seafaring men, too, they were well regarded, for
your sailorman of a past generation knew when the
commander and officers of a ship were thoroughly profi-
cient as seamen, and under their house flag none other
served. Aboard their vessels, men shipped voyage after
voyage, for "Belaying Pin Soup," "Knuckle Duster"

164

warfare and other too prevalent forms of human ill-treatment were not countenanced.

For a figurehead the *Flying Fish* carried a fish on the wing of lifelike color and giving a vivid sense of speed. She was one of the fastest and most beautiful clippers that ever sailed under the Stars and Stripes. She measured: length 198 feet, 6 inches, breadth 38 feet 2 inches, depth 22 feet, with 25 inches dead-rise at half floor.

She became noted as a contestant in more famous ocean races than any one of McKay's many clippers. On her maiden voyage it was arranged that she should race the *Swordfish* around the Horn to San Francisco.

In the principal hotels of New York, especially the Astor House, ship-owners and masters, brokers and others connected with shipping, would foregather and discuss the merits of Webb's *Swordfish* and McKay's *Flying Fish*. Each of the clippers had her devoted admirers and heavy bets were made as to the relative speed and the length of their passage to San Francisco. The records and "driving" ability of each ship's commander, also, would enter into the discussion. There was considerable rivalry between the partisans of William H. Webb, New York's leading clipper shipbuilder, and Donald McKay. Comparison of passages made by the ships of each builder, would be made, and, too frequently, cause ill-feeling, expecially if contestants represented Boston versus New York.

These two extreme clippers were to sail around the Horn, over the great ocean racecourse, fifteen thousand miles in length, where some of the most glorious trials of speed and of prowess that the world ever witnessed have taken place. Here the American clipper ship of the fifties, the noblest work that has ever come from the hands of man, was sent, to contend with the elements, to outstrip steam, and astonish the world!

The *Flying Fish* sailed from Boston November 11, 1851, and on the same day the *Swordfish* passed Sandy Hook, Captain Nickels of the *Flying Fish* and Captain Babcock of the *Swordfish* were both young and skilful commanders, and it was believed by their friends that each would send his ship along at her utmost speed. The *Flying Fish* made an excellent run of 19 days to the equator, leading the *Swordfish* by four days. From the equator to 50° S., the *Flying Fish* was 26 and the *Swordfish* 22 days, so that they passed that parallel on the same day. They raced round Cape Horn, part of the time side by side, the *Flying Fish* making the run from 50° S. in the Atlantic to 50° S. in the Pacific in 7 and the *Swordfish* in 8 days. From this point the *Swordfish* came up and steadily drew away. She made the run to the equator in 19 days, leading the *Flying Fish* by 3 days, and from the equator to San Francisco in 20 days, gaining on this stretch another 3 days, and arrived at San Francisco, February 10, 1852, after a splendid passage of 90 days

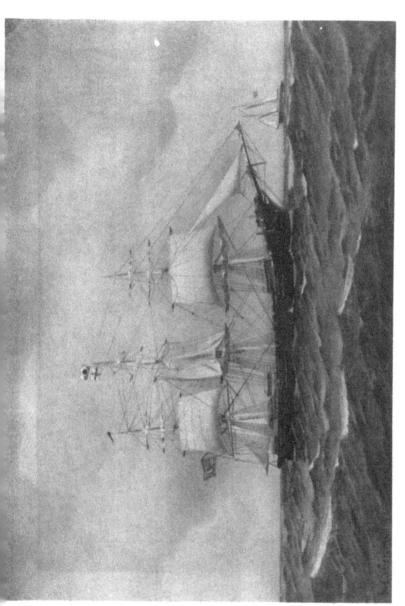

"FLYING FISH" LAYING TO FOR A PILOT

From painting by Fred S. Cozzens in possession of the author.

THE CLIPPER RACE.—"FLYING FISH" IN THE LEAD!

16 hours from New York. The *Flying Fish* arrived on the 17th, or 98 days from Boston.

From San Francisco the *Flying Fish* sailed to China, but like most of McKay's powerful sailers she did better on the California run. She reached New York, however, in time to become a contestant in the most celebrated and famous race that has ever been run off upon a trackless waste of ocean. Yes, and she won against one of the finest fleets of clippers that ever sailed the seas!

The astounding discovery of the treasure buried beneath the soil of the long-considered barren desert of California, created an unwonted stir amongst the shipping in the port of New York, as well as in every other seaport of the United States; for every merchant, who could by any means manage to do so, was eager to possess himself of a share of the almost fabulous wealth then for the first time taking hold on men's imaginations. Consequently all vessels that could be obtained, were readily freighted for the shores of the "land of golden promise." Freights increased in value—there was a rise in seamen's wages—a scarcity of seamen for every other destination, and a proportionate increase in everything connected with shipping.

In order to convey to the reader an idea of shipping conditions then and how sailors were shipped for long off-shore voyages, we believe the accompanying description of a scene in a New York shipping office serves the purpose:

One fine afternoon in the latter part of October, 1852, there was an unusual bustle in a well-known shipping office for seamen, in the vicinity of the East River, for the proprietor had within a few days received almost unlimited requisitions from divers ship-owners and masters, to procure crews for their vessels. The office was thronged with seamen, "outward bounders," as, in nautical parlance, Jacks were termed, when they had run their tethers to their full length ashore—in other words, when they had spent their hard-earned wages of months, in a few days, and found themselves snubbed by the landlady and shunned by the fair, but frail nymphs, in the sunshine of whose favors they had so lately basked.

Strange to say, there was not that seeming anxiety amongst the men to get a berth, almost without regard to the place of their destination, that was customary when the throng was so great. The hardy fellows rather seemed to hang back—although they had not "a shot left in the locker"—and to feel a desire to pick and choose with all the nicety and independence of a cautious "homeward bounder" who takes warning from past follies and mishaps, and goes to sea again before his pockets had been drained by land sharks in the shape of boarding-house and saloon keepers, sharpers, alongshore-men, etc. A stout, burly, black-whiskered, weather-beaten looking man walked into the office, elbowed his way through the crowd of sailors, and approaching the shipping master, said, "Good morning, Mr. Sharply—have you got my crew all ready? We haul out at four o'clock this afternoon, and I must have the men all on board by that time."

"I am sorry to say, Captain," replied the shipping master, "that there are six hands wanting yet. 'Pon my word, in all my experience I never saw such difficulty in procuring sailors for these northern voyages." Then, raising his voice, and addressing the assembled seamen, he shouted—

"Now, lads, who's for a trip to Antwerp in the *Mermaid?* Fine ship—A-1—good captain—first rate provisions—tiptop wages, and a capital time of year for the voyage. Six hands wanted, at two dollars a month above ordinary wages." There was no reply to these tempting offers.

The captain and the shipping master looked disappointed, but in the course of a few minutes, the former left the office, saying "sotto voce" to the shipping master, as he turned to go away—

"Mr. Sharply, you must get me six men by four o'clock, by hook or by crook—some way or other; even if you have still to increase the wages."

Scarcely had the captain of the Antwerp ship left the office before another captain entered. He saluted the shipping master familiarly, and asked if his crew was ready.

"Hardly yet," replied the latter smiling, "hardly yet, Captain Nickels. I only received your order an hour ago—but I shall have no difficulty in procuring you a first-rate crew immediately."

"Who wants to go to San Francisco," continued he, addressing the sailors, "I want fourteen hands for the McKay clipper *Flying Fish* to sail tomorrow morning?"

Scarcely had the words escaped his lips, before there was a rush to the desk of twice the number of hands wanted, all eager to register their names for the voyage. The requisite crew was soon procured, for the *Flying Fish's* second ocean race to 'Frisco.

An account of it has been written by Lieutenant M. F. Maury, than whom no man ever presented the world with such beneficial results for the improvement of commerce and navigation. Navigators were beginning fully to reap the benefit of Maury's researches, when four splendid American clipper ships put to sea from New York—all bound for California.

They were ably commanded, and, as they passed the bar at Sandy Hook, one by one, and at various intervals of time, they presented really a most magnificent spectacle. The names of these ships and their masters were, the *Wild Pigeon*, Captain Putnam; the *John Gilpin*, Captain Doane; the *Flying Fish*, Captain Nickels, and the *Trade Wind*, Captain Webber. Like steeds that know their riders, they were handled with the most exquisite skill and judgment, and in such hands they bounded out upon the "glad waters" most gracefully. Each, being put upon her mettle from the start, was driven, under the seaman's whip and spur, at full speed over a course that it would take them three long months to run.

The *Wild Pigeon* sailed October 12; the *John Gilpin*, October 29; the *Flying Fish*, November 1; and the *Trade Wind*, November 14. All ran against time; but the *John Gilpin* and the *Flying Fish* for the whole course, and the *Wild Pigeon* for part of it, ran neck and neck, the one against the other, and each against all. It was a sweepstake with these ships around Cape Horn and through both hemispheres.

Lieutenant Maury wrote:

Wild Pigeon led the other two out of New York, the one by seventeen, the other by twenty days. But luck and chances of the winds seem to have been against her from the start. As soon as she had taken her departure, she fell into a streak of baffling winds, and then into a gale, which she fought against and contended with for a week, making but little progress the while; she then had a time of it in crossing the horse latitudes. After having been nineteen days out, she had logged no less than thirteen of them as days of calms and baffling winds; these had brought her no farther on her way than the parallel of 26° north in the Atlantic. Thence she had a fine run to the equator, crossing it between 33° and 34° west, the thirty-second day out. She was unavoidably forced to cross it so far west; for only two days before, she crossed 5° north in 30°—an excellent position.

So far, therefore, chances had turned up against the *Pigeon*, in spite of the skill displayed by Putnam as a navigator, for the *Gilpin* and the *Fish* came booming along, not under better management, indeed, but with a better run of luck and fairer courses before them. In this stretch they gained upon her—the *Gilpin* seven and the *Fish* ten days; so that now the abstract logs show the *Pigeon* to be but ten days ahead.

Evidently the *Fish* was most confident that she had the heels of her competitors; she felt her strength, and was proud of it; she was most anxious for a quick run, and eager withal for a trial. She dashed down southwardly from Sandy Hook, looking occasionally at the Charts; but, feeling strong in her sweep of wing, and trusting confidently in the judgment of her master, she kept, on the average, two hundred miles to leeward of the right track. Rejoicing in her many noble and fine qualities, she crowded on her canvas to its utmost stretch, trusting quite as much to her heels as to the Charts, and performed the extraordinary feat of crossing, the sixteenth day out from New York, the parallel of 5° north.

The next day she was well south of 4° north, and in the doldrums, longitude 34° west.

Now her heels became paralyzed, for Fortune seems to have deserted her a while—at least her master, as the winds failed him, feared so; they gave him his motive power; they were fickle, and he was helplessly baffled by them. The bugbear of a northwest current off Cape St. Roque began to loom up in his imagination, and to look alarming; then the dread of falling to leeward came upon him; chances and luck seemed to conspire against him, and the mere possibility of finding his fine ship back-strapped filled the mind of Nickels with evil forebodings, and shook his faith in his guide. He doubted the Charts, and committed the mistake of the passage.

The Sailing Directions had cautioned the navigator, again and again, not to attempt to fan along to the eastward in the equatorial doldrums; for, by so doing, he would himself engage in a fruitless strife with baffling airs, sometimes re-enforced in their weakness by westerly currents. But the winds had failed, and so too, the smart captain of the *Flying Fish* evidently thought, had the Sailing Directions. They advise the navigator, in all such cases, to dash right across this calm streak, stand boldly on, take advantage of slants in the wind, and, by this device, make easting enough to clear the land. So, forgetting that the Charts were founded on experience of great numbers who had gone before him, Nickels, being tempted, turned a deaf ear to the caution, and flung away three whole days, and more, of most precious time, dallying in the doldrums.

He spent four days about the parallel of 3° north, and his ship left the doldrums, after this waste of time, nearly upon the same meridian at which she entered them.

She was still in 34°, the current keeping her back just as fast as she could fan east. After so great a loss, her very clever master, doubting his own judgment, became sensible of his error. Leaving the spell-bound calms behind him, where he had undergone such trials, he wrote in his log as follows: "I now regret that, after making so fine a run to 5° north, I did not dash on, and work my way to windward to the northward of St. Roque, as I have experienced little or no westerly set since passing the equator, while three or four days have been lost in working to the eastward, between the latitude of 5° and 3° north, against a strong westerly set;" and he might have added, "with little or no wind."

The *Wild Pigeon*, crossing the equator also in 33°, had passed along there ten days before, as did also the *Trade Wind* twelve days after. The latter also crossed the line to the west of 34°, and in four days after had cleared St. Roque.

But, notwithstanding this loss of three days by the *Fish*, who so regretted it, and who afterward so handsomely retrieved it, she found herself, on the 24th of November, alongside of the *Gilpin*, her competitor. They were then both on the parallel of 5° south, the *Gilpin* being thirty-seven miles to the eastward, and of course in a better position, for the *Fish* had yet to take advantage of slants, and stand off shore to clear the land. They had not seen each other.

The Charts showed the *Gilpin* now to be in the best position, and the subsequent events proved the Charts to be right, for thence to 53° south the *Gilpin* gained on the *Pigeon* two days, and the *Pigeon* on the *Fish* one.

By dashing through the Straits of Le Maire, the *Fish* gained three days on the *Gilpin;* but here Fortune again deserted the *Pigeon*, or rather the winds turned against her; for as she appeared upon the parallel of Cape Horn, and was about to double round, a westerly gale struck her and kept her at bay for ten days, making little or no way, except alternately fighting in a calm or buffeting with a gale, while her pursuers were coming up "hand over fist," with fine winds and flowing sheets.

They finally overtook her, bringing along with them propitious gales, when all three swept past the Cape, and crossed the parallel of 51° south on the other side of the "Horn," the *Fish* and the *Pigeon* one day each ahead of the *Gilpin*.

The *Pigeon* was now, according to the Charts, in the best position, the *Gilpin* next, and the *Fish* last; but all were doing well.

From this parallel to the southeast trades of the Pacific the prevailing winds are from the northwest. The position of the *Fish*, therefore, did not seem as good as the others, because she did not have the sea-room in case of an obstinate northwest gale.

But the winds favored her. On the 30th of December the three ships crossed the parallel of 35° south, the *Fish* recognizing the *Pigeon;* the *Pigeon* saw only a "clipper ship," for she could not conceive how the ship in sight could possibly be the *Flying Fish*, as that vessel was not to leave New York for some three weeks after she did; the *Gilpin* was only thirty or forty miles off at the same time.

The race was now wing and wing, and had become exciting. With fair winds and an open sea, the competitors had now a clear stretch to the equator of two thousand five hundred miles before them.

The *Flying Fish* led the way, the *Wild Pigeon* pressing her hard, and both dropping the *Gilpin* quite rapidly, who was edging off to the westward.

The two foremost reached the equator on the 13th of January, the *Fish* leading just twenty-five miles in latitude, and crossing in 112° 17'; the *Pigeon* forty miles farther to the east. At this time the *John Gilpin* had dropped two hundred and sixty miles astern, and had sagged off several degrees to the westward.

Here Putnam, of the *Pigeon*, again displayed his tact as a navigator, and again the fickle winds deceived him: the belt of northeast trades had yet to be passed; it was winter; and, by crossing where she did, she would have an opportunity of making a fair wind of them, without being much to the west of her port when she would lose them. Could she have imagined that, in consequence of this difference of forty miles in the crossing of the equator, and of the two hours' time behind her competitor, she would fall into a streak of wind which would enable the *Fish* to lead her into port one whole week? Certainly it was nothing but what sailors call "a streak of ill luck" that could have made such a difference.

But by this time *John Gilpin* had got his mettle up again. He crossed the line in 116°—exactly two days after the other two—and made the glorious run of fifteen days thence to the pilot grounds of San Francisco.

Thus end the abstract logs of this exciting race and these remarkable passages.

The *Flying Fish* beat; she made the passage in 92 days and 4 hours from port to anchor; the *Gilpin* in 93 days and 20 hours from port to pilot; the *Wild Pigeon* had 118. The *Trade Wind* followed, with 102 days, having taken fire, and burned for eight hours on the way.

On her next passage, Boston to San Francisco, the *Flying Fish* took 109 days, and late in the same year, 1855, she made another run in 105 days. She was afterwards engaged in various trades until 1857, when she made her last trip around the Horn, arriving at 'Frisco in 100 days from Boston.

While coming out of Foochow in November, 1858, with a cargo of tea, she was wrecked, and then abandoned to the underwriters, and the wreckage sold to a Spanish merchant of Manila. She was subsequently floated and rebuilt at Whampoa, her name being changed to *El Bueno Sucesco*. She sailed for some years between Manila and Cadiz—and finally foundered in the China Sea.

CHAPTER XXII

DONALD McKAY was not satisfied with the speed of the *Flying Cloud*, which on her first passage, not only made the quickest run from New York to San Francisco then known, but attained the highest rate of speed on record. Such results would have satisfied most men that they had at last produced a model which might defy competition, and would have flattered themselves during the rest of their lives, that they had discovered perfection, and they would have made no further efforts to excel. But such were not the conclusions of the designer of the *Flying Cloud*. His daring and active mind was not satisfied with a single triumph. He carefully reviewed all his past works and analyzed their results, and came to the conclusion that perfection in modelling had not yet been discovered; that whatever success he had attained had been the result of increased information, derived from experience, and that if he desired to excel he must never cease to improve. Impressed with these views, he determined to build a clipper that should outsail *Flying Cloud*, and he produced the *Sovereign of the Seas*.

177

More than two centuries had passed away since this name was first applied to a ship. In 1637, her patronymic *Soveraigne of the Seas*, was built at Woolwich Dock Yard, England, in the building of which this interesting description is given—"Her burthen was just as many tons as there had been years since our Blessed Saviour's Incarnation, viz.—Sixteen hundred and thirty-seven, and not one under or over." She was the first ship with "flushe deckes" and the largest which had previously belonged to the English Navy,—so Mr. McKay could not have selected a better name for his own-ship; its historical association is full of instruction, and no ship was ever more worthy of such a name.

The *Sovereign of the Seas* was said to have originally been named *Enoch Train*, a merited compliment to one of the most enterprising merchants of his time—a gentleman who did so much for the shipping interests of Boston, and always remained the true friend and staunch supporter of Donald McKay. Train & Co. contemplated purchasing this vessel when she was upon the stocks, perhaps the following attributed reason for the change in name, from George Francis Train's biography, is not without some foundation:

The building of the *Flying Cloud* was a tremendous leap forward in shipbuilding; but I was not satisfied, [said Train]. I told McKay that I wanted a still larger ship. He said that he could build it. And so we began another ship that was to outstrip in size and capacity the great *Flying Cloud*. I was desirous to name this ship the *Enoch Train*, in honor of the head of the Boston house, and had said as much to Duncan McLean, who was the marine reporter for the *Boston Post*. McLean had usually written a column for his paper on the launching of our ships. He wanted to have something to write about the new vessel. I told him the story of Colonel Train's life, and that we were going to christen the new vessel with his name. I did not consult Colonel Train, thinking that, of course, it was all right. The *Post* published a long account of the ship, and gave the name as the *Enoch Train*. When I went down to the office that morning Colonel Train had not arrived, but soon came in, walking straight as a gun-barrel, and seeming to be a little stiff. 'Did you see the *Post* this morning?' I asked. 'Premature,' he replied. That was all he said. He would not discuss the matter. I was nettled that he did not appreciate the honor I thought I was con-ferring on him. It was not for nothing that a man's name should be borne by the greatest vessel on the seas. I said to myself that the name should be changed at once. The ship was to be of 2200 tons' burden and I decided to call her the *Sovereign of the Seas*.

Amidst his extensive shipbuilding operations, late in the year 1851, Donald McKay was notified by the city authorities that the land occupied by his shipyard would be taken over for an extension of Border Street, in East Boston, so he had established himself in another location a short distance north of his former yard. It was at this new yard that many thousand people assembled to view the launching of his magnificent clipper one Saturday afternoon, in July, 1852. She was gaily decorated with flags and looked beautiful, as she glided along the ways into the sea, saluted with cheers upon cheers. Bowing obeisance to the spectators on the shore, she went gracefully about twice her length and grounded. Next tide she was towed off, taken to Lombard's Wharf, East Boston, and rigged with despatch.

The *Sovereign of the Seas* was pronounced "the longest, sharpest, and most beautiful merchant ship in the world," designed to sail, at least 20 miles an hour in a whole-sail breeze. Her extreme length was 265 feet, breadth of beam 44 feet, length of Keel 245 feet, length between perpendiculars 258 feet, breadth of gunwales 42 feet, depth of hold 23½ feet, including 8 feet of 'tween decks. It was estimated she could stow nearly 3,000 tons of measurement goods and not draw more than 20 feet of water. Her admirable proportions, her graceful curved lines, her lofty masts, and her rigging, of unsurpassed completeness, her strength of construction and beauty of interior finish; all that met the eye impressed and held in delighted wonder the visitor to this queenly Boston clipper.

Donald McKay, building this ship on his own account, of course aimed to realize his ideal of a beautiful vessel. Probably no one but a shipbuilder can fully sympathize with the triumphant feeling which the sight of such a piece of workmanship inspires in the breast of him who has planned its mighty proportions, and watched the execution of the whole design. And perhaps no one but a Captain can understand the feelings of the master of such a ship, out upon the wide ocean, with such a power and such a responsibility committed to him. A man standing on her quarter-deck in command running down the trades with a whole-sail breeze, came as near being a "sovereign of the seas" as a man very well could.

As you stepped on board, the first impulse would be to look aloft. The sight was grand. The mainmast rose 92¾ feet, and the whole system of masts, yards, and rigging loomed so large and loftily that it did not seem designed to be managed by human hands, arrested the gaze with peculiar power. In a single suit of sails she spread about twelve thousand yards of canvas! Her mastheads were covered with gilded balls, her yards black, booms bright and lower masts white, and altogether aloft, she was the best-fitted ship then ever built in Boston. Her yards were all of single spars, and together with the masts, strong enough to stand in a gale till every stitch of canvas blew away. She had the stoutest and most beautifully proportioned set of spars imaginable. The bowsprit, made of hard pine, was 20 feet out-board and 34 inches in diameter.

Gracefully ornamenting her bow, was the appropriate figure of a sea god, half-man and half-fish,—Neptune, good old Neptune, with a conch shell to his mouth, as though he were blowing it. In stowage capacity, strength of construction, and beauty of outline, she ranked the foremost of the American clipper fleet.

Contrary to the advice of his best friends, Donald McKay had built the *Sovereign of the Seas* upon his own account; he embarked all he was worth in her, for no merchant in New England would risk capital in such a vessel, as she was considered too large and costly for any trade. But his clear mind saw the end from the beginning. Before her keel was laid, he had mastered the workings of the California trade, and when she was ready, to the surprise of even those who knew him best, he played the merchant successfully and loaded her himself. His business capacity for mercantile transactions was scarcely less conspicuous than his skill as a mechanic. In command of his ship, he placed his brother, Lauchlan McKay, who, besides having been a master in the trans-Atlantic packet service, was a practical shipbuilder and in every way competent for the position.

Upon arrival at New York the *Sovereign* attracted much attention among those interested in naval architecture.

Messrs. Grinnell, Minturn & Co. were so justly proud of being the consignees of so splendid a craft that they gave a dinner on board to some of New York's prominent merchants and shipping men. Below is a report of the "proceedings" taken from a newspaper of that time:

The two spacious and elegant saloons were comfortably filled, Moses H. Grinnell presiding in one, Captain McKay in the other, assisted by George W. Blunt, Esq. The roominess and elegance of the ship seemed to have infused a universal cheerfulness and much pleasantry prevailed. Mr. Blunt made a capital speech in proposing the health of Captain McKay, who responded with the true modesty of innate worth, and very happily proposed the health of Mr. Blunt.

Captain Nye's pleasant face, always turned to the Pacific, now loomed up to leeward of a chowder tureen, and his manly, cheery voice was heard hailing the "Hon. Daniel Webster, the Lion of the Nation." The toast being recognized as not now political, Mr. Grinnell was hailed for a response. That merchant prince, a nobleman of the right mould, soon appeared in the "Golden Gate" that separated the two saloons, and bearing up to the windward, took Mr. Blunt's chair instead of his chart, and after a little backing and filling, paid an eloquent tribute to the great statesman's merits, reasserting his claims upon the country and especially upon the mercantile community, but declaring himself now to be in the ranks of General Scott's army. Then followed sundry pleasantries about the candidates for the Presidency, and the very pleasant company broke up, first joining in a unanimous "Success to the *Sovereign of the Seas*—our only Sovereign."

Donald McKay's ship was loaded with the largest cargo ever despatched from the port of New York, amounting to about 2950 tons of assorted merchandise, exclusive of stores for a year's voyage; and, also, probably with a larger freight list than ever before cleared from New York. She was despatched by her agents, Messrs. Grinnell, Minturn & Co., in thirty working days.

The following is a list of her passengers:—

Mrs. Charles Stout and son, C. A. Poor, Boston; H. F.

Poor, do; B. Byrne, Jr., W. L. Ryckman, Mrs. G. W. Ryckman, and two children; D. T. Moore, D. T. Louchk, J. T. Van Duzen, W. D. Hees, Mrs. R. Gorham, six children and servant.

This may interest some of our readers, although it stands no comparison with the passenger list of one of our modern leviathans.

And now we can imagine the clipper, *Sovereign of the Seas* ready "to walk the waters like a thing of life." She sailed from New York, August 4th, with a fine leading breeze, but during the night, the wind changed ahead and blew a gale. The noble ship, however, clawed off shore like a pilot boat, carrying whole topsails, courses, jib and spanker. Next morning, the wind favored her a little, and she was soon under all sail, close-hauled, walking to the eastward at the rate of 15 miles an hour, and long before sunset she was out of sight of land.

Here let us give some idea of the discipline of the ship. Captain McKay had a picked crew, including officers, of 103 men and boys, consisting of

Mates.............	4
Boatswains.........	2
Carpenters.........	2
Stewards...........	3
Cooks.............	2
Able bodied seamen	80
Boys...............	10
Total...........	103

These were divided into two watches and stationed man-of-war fashion, on the forecastle, in the tops, in the waist, and on the quarter deck; so that in working ship, making or taking in sail every man was at his station. All these arrangements were made by the captain, himself. He stationed his men and laid down the rules which should govern both him and them during the passage,— everything was done quietly and orderly.

It is related that the chief mate, a big swaggering bully, began swearing at the men. This the captain checked at once, not in the presence of the men, but privately. His reproof, however, was so mild and gentlemanly, that the mate thought Captain McKay afraid of him and persisted with increased insolence, highly seasoned with "Dutch courage." He even went so far as to countermand the Captain's orders and mutinously attempted to assume command of the ship. Without any explosion of wrath Captain McKay ordered him off duty, when twenty days out. His services were not again required during the passage.

The first sixteen days, the *Sovereign of the Seas* encountered strong southerly winds, and made only 600 miles southing. Twenty-seven days out she crossed the Equator, having worked almost every inch of the way dead to windward. When she reached the Falkland Islands, tremendous S. W. gales with a sea awful to look at were encountered, yet her Captain carried a heavy press of sail, and actually drove his ship through, standing off and on every four hours. Hail, snow and screamers were the companions of the crew day and night, but most nobly did the ship behave.

And here let us show what kind of a man Lauchlan McKay was to his crew. He had stoves in their quarters, and continually had one or more of the boys attend the fires and at the same time dry the sailors' clothes; warm coffee and tea, and provisions were served out during the night, as well as the day, and he never exposed his men more than absolutely necessary. Still he carried on sail so as to make it truthfully frightful to look aloft, and fairly beat his ship dead to windward, against head gales and currents from the Falkland Islands to Cape Horn, a feat, we believe, never before performed, under such circumstances; and what redounds to his credit, is the fact, that by his judicious conduct, not a man was made sick or disabled. The *Sovereign of the Seas* doubled the Cape in 51 days, and here, by the way of change, had four days calm.

From the Cape she had head winds, calms, and gales by turn, and it seems astonishing how she ever got along. On the night of October 12th, during a heavy gale, but carrying as usual a press of canvas, the main topmast trestle-trees settled, which slackened the topmast back-stays, and away went the maintopmast over the side,

186

"SOVEREIGN OF THE SEAS" DISMASTED OFF VALPARAISO.

From a picture in the author's possession.

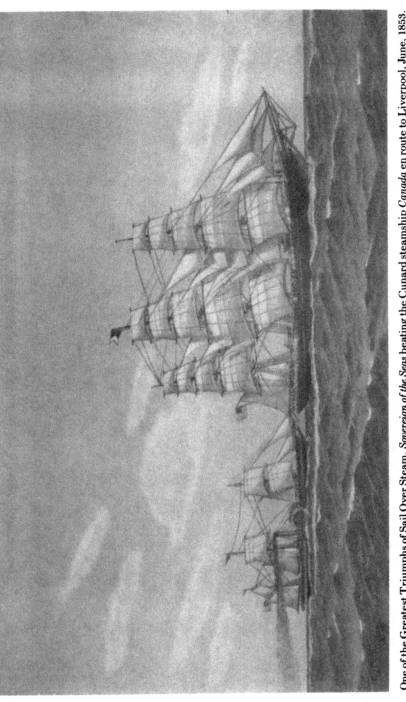

One of the Greatest Triumphs of Sail Over Steam. *Sovereign of the Seas* beating the Cunard steamship *Canada* en route to Liverpool, June, 1853. From painting by Fred S. Cozzens in possession of the author.

taking with it the foretopmast, foreyard and mizzen topgallantmast, and every stitch of canvas off the foremast. Here was a disaster to make the boldest heart beat quick, and even palsy the tongue!

The hands were called, the ship hove to; and, now, said Captain McKay to the second mate (acting mate), "You take the mainmast, and I will take the foremast, and let us clear the wreck. Remember, everything must be saved—nothing must be cut."

"Impossible, sir," replied the mate, "we must cut the wreck adrift." "I repeat," said the Captain, in a tone of voice not to be mistaken, "nothing shall be cut;" and turning familiarly to the crew, said—"Boys, nothing is impossible to him that wills! I will that everything shall be saved. Now go to work like Trojans." And to work they went in earnest. They vied with each other in going overboard to clear the wreck—not a murmur was heard fore or aft, and before sunset the next day everything was on board, and the ship under her mainsail, crossjack course and mizzen-topsail, was balling off 12 knots. Her decks were lumbered up to the leading blocks. The Captain was everywhere; now setting a sailmakers' gang to work repairing sails, next a carpenter's gang to making and fitting masts and yards, and the sailors generally to clearing the rigging, and getting down the stumps of the topmasts. Every man was employed and worked with a will, but at night the watch was regularly set, though Captain McKay himself did not sleep. The watch on deck worked during the night and all hands during the day. In a week both topmasts, topsail yards and fore yard were aloft and the sails bent, and in 12 days the ship was once more a-tanto, and as complete aloft as if nothing had happened. Captain Lauchlan McKay's skill as a sailor, his dauntless energy as a man, his kindness to his crew, and his entire abnegation of self, all stamped him as a truly great commander. His brother Donald's confidence in him had been vindicated."

The foregoing account is taken from a statement by one of the *Sovereign's* crew. It proves beyond a peradventure, that all American crews were not subject to the excessively brutal treatment which certain writers, especially foreigners, delight in narrating with exaggeration.

This gallant ship arrived at San Francisco, November 15th, 1852, in 103 days, having beaten, it is claimed, every vessel that sailed within a month of her.

When she was hauling into the wharf, the whole shore within sight of her was lined with people, who cheered her again and again. Merrily the sailors sang, as they whirled her capstan 'round:

> "O, Susannah, darling, take your ease
> For we have beat the clipper fleet,
> The *Sovereign of the Seas!*"

Many stories have been told of the speed with which, during the gold fever in California, the crews of incoming vessels would jump from the decks and run for the mines. The strongest contracts made in New York would not hold them. Captain McKay must have been much liked by his men, or they would not have stayed and moored his ship snugly.

When she arrived at 'Frisco, flour was selling at $44 per barrel! And her freight and passage money amounted to over $98,000! A fortune in itself and which is apt to cause the present day American ship-owner to groan aloud!

Instead of running across to China, as the clippers often did to reload home for the Atlantic Coast, Captain McKay crossed to Honolulu and there secured a cargo of whale oil taken on the Pacific grounds by whaling ships from New England. Thus the *Sovereign of the Seas* created what was a boom for the whaling industry, by opening up cheap transportation of their catch from this rendezvous in the Sandwich Islands to the whalers' home ports. Besides being saved all expense incident to the heretofore necessary long voyage, whalers were enabled to continue cruising, so that this innovation proved of benefit to all concerned.

From the files of a Polynesian newspaper, we learn that she was visited by His Majesty and suite, accompanied by the American and other Consuls, together with a number of residents, who spent an hour or two on board, all of whom were highly pleased with the ship and the attentions they received from Captain McKay. Besides some passengers, the *Sovereign* had taken on board at San Francisco, one large grizzly bear, a rainbow bear, a wolf, a coyote, a wildcat and a leopard, all destined for exhibition in the Crystal Palace, New York, which unusual cargo must have surprised and delighted the Sandwich Islanders.

With a crew of only 34 men (her previous complement,

New York to Frisco, had been 103), and the additional drawback of crippled fore and main topmasts, the *Sovereign* obtained the highest rate of speed on record. In 22 days she ran 6245 statute miles (one-fourth the distance around the earth), and during 11 of these days consecutively, her daily average was 354 statute miles; and during four days, also consecutively, she averaged as high as 394¾ statute miles,—a greater distance than was ever before steadily logged for a twenty-four hour run by any craft afloat. Her homeward passage from Honolulu to New York occupied 82 days—a remarkably short run!

> There is a tide in the affairs of men
> Which taken at the flood, leads on to fortune.

And now let us revert to the master-builder who, daringly confident in the product of his own labor and skill staked his whole fortune and reputation in a venture that was considered almost foolhardy. It must be remembered, he had been compelled to play the merchant. He was well rewarded, for the actual earnings of his ship for nine months (say from August, 1852, to May, 1853) amounted to about $135,000—a handsome return for her first round voyage; and he soon afterward sold her upon his own terms!

Upon his arrival at New York, Captain Lauchlan McKay was presented with a valuable service of silver plate, by the Insurance Underwriters, as a token of their appreciation of his skill during the voyage to California and back.

Mr. Walter Jones in behalf of the insurance interests, spoke, as follows:

Capt. McKay: The perseverance and skill you displayed in refitting your ship at sea, when dismasted, and continuing your voyage without expense to the underwriters, is worthy of high commendation, and is a guarantee for the future that all a man can do in any emergency will be done by you. Accept, therefore, dear sir, this, a token of our esteem, and long may you live to use it.

Capt. McKay replied:

Mr. Jones and gentlemen—I gratefully accept your generous present, valuable in itself, but more so for the sentiments of approbation with which you have been pleased to associate its presentation. I thank you, too, in behalf of my gallant crew, for the eight hundred dollars which you gave them. In the hour of trial they were more precious than gold, and to their untiring devotion I am indebted for whatever credit I have received. We all endeavored to do our duty, and the only precedence I am entitled to was the result of my position, not, I assure you, on account of extra merit. I shall, therefore, ever regard this token of your approbation, as shared by my crew, whose services will be associated with it in my remembrance while I live.

In May, 1853, a Clipper challenge in relation to the *Sovereign* caused some excitement among Boston and New York ship-owners. It was not only a contest between the shipping interests of both cities, but the outcome of keen rivalry between Donald McKay and William H.

Webb. California freights were very dull and all kinds of schemes were concocted to procure shipments. With this object in view, a party interested in loading the *Young America* (then recently launched from Webb's yard), inserted a challenge in one of the New York papers that the *Sovereign of the Seas* would race against any ship in the world to San Francisco for $10,000. A short time previously Mr. McKay had sold the *Sovereign* to Funch & Meincke, well-known ship brokers of New York. The impression went abroad that this challenge was given by the new owners, yet nothing could be more erroneous, for at that time they were planning a voyage to Liverpool. Mr. Webb prudently ventured to name $10,000. The owners of the *Sovereign*, to show their pluck, stumped Webb by stating that when she returned from England they would back her sailing qualities against those of any ship Webb ever built, but their ship went to Australia instead of returning to this country from England.

On June 18, 1853, she sailed from New York to Liverpool. As California freights had been falling and the China trade was dull at the time, she had been loaded for Liverpool, where, later on, she was chartered by James Baines & Co. for their Australian trade.

In order to study the behavior of his vessels at sea and the various considerations that should govern form of hull and arrangements of the spars and sails, for he was a man that gave his whole attention to the art of shipbuilding, Donald McKay crossed the Atlantic on the *Sovereign* and probably the result of observations he made upon this passage had much to do with the superior excellence of his later vessels. To keep employed the force of skilful and well-trained men he had gathered together and the capital he had accumulated must have been a herculean task; his ingenuity must have been severely taxed to devise improvements in models and rig, and we daresay he was compelled to exercise sleepless care in regard to materials and the manner of putting them together.

On this voyage Donald McKay was accompanied by his wife. He had been very fortunate in his second marriage, for he found a woman of sterling character, loyal to him and his family throughout the many years of their life together, Mary Cressy Litchfield,—the daughter of Nichols and Anna Cushing Litchfield, who had lived at East Boston for some years. Called upon to assume charge of his large household, as well as the many needs of a growing family, this very capable young woman showed her ability to govern from the start. She was of invaluable assistance to her husband in the conduct of his business affairs, notwithstanding all the home cares devolving upon her. As the romantic names of so many of his ships are worthy of notice and have been attributed to Donald McKay, peculiarly happy in his choice of fitting titles, we believe due credit should now be given

Mary Cressy McKay, as the real originator of their attractive and befitting nomenclature. She survived her famous husband many years, until February 6, 1923,—a truly wonderful woman who outlived her own generation to be loved and appreciated by the next.

The *Sovereign of the Seas* still remained true to her aspiring name, for on this run she eclipsed all previous trans-Atlantic crossings—13 days, 22 hours, 50 minutes, from dock in New York to anchorage in the Mersey; she sailed from the Banks of Newfoundland to Liverpool in 5 days, 17 hours.

The Cunard Steamship *Canada* sailed from Boston on the same day that the *Sovereign* left New York. By comparing her log with that of this steamer, it was found that the ship on June 25, 1853, was ahead 240 miles. On the 30th, she had beaten the steamer 325 miles. The steamer's greatest day's run was 306 miles and the same day the ship logged 340 miles; and the ship, too, was drawing 22 feet water, and rather crank, having been badly laden.

The *Sovereign's* time, it is stated by competent authority, has never been equalled by a sailing vessel in the month of June. Her best day's run was 344 miles, June 28th, the log reading—"Wind N.—Took in topgallant sails, single-reefed topsails; ship very crank, lee rail under water; rigging slack, weather squally, ship rolling heavily, strong breezes."

Three-quarters of a century afterward, it is enlivening to read about the wonderful performances of this splendid ship

Some Englishmen were Donald McKay's guests at East Boston, representing James Baines' and other British shipping interests. He cordially invited them to accompany him, as they were returning to England at the same time his clipper was scheduled to sail. They refused with haughty disdain, saying that their preference was a Cunard steamship,—that time was pressing and they desired to get to Liverpool quickly, no sailing ship could reach there in time, etc. Mr. McKay was nettled and he made up his mind to beat them across the Atlantic. And he did! When the *Canada* arrived at Queenstown Donald McKay's clipper had passed there several hours previously and conspicuously spread aloft when she docked at Liverpool was a large canvas sign; reading,

"SOVEREIGN OF THE SEAS,"
FASTEST SHIP IN THE WORLD—
SAILED NEW YORK TO LIVERPOOL
RECORD TIME—13 DAYS, 22 HOURS."

At Liverpool, the *Sovereign* attracted much attention and was chartered by James Baines & Co. for a voyage to Australia, in their line known as the "Black Ball of Liverpool." (Why James Baines ever took this name for his fleet never has been satisfactorily explained. It caused considerable confusion in shipping circles and the old-established trans-Atlantic American Packet line similarly named, requested him to adopt another but without success.)

Meantime, the *Sovereign's* short sailing passage from New York to Liverpool, caused no little discussion. Among others, Captain E. Nye, who had commanded the American packet *Independence* entered the controversy with vehemence. He was silenced when challenged to produce his log against McKay's flyer. British shipping men were rather jealous of the fact that this fine American ship had made the trans-Atlantic passage in less time than ever previously accomplished, and one writer in the London *Times*, by way of throwing it into the shade, stated that the British frigate *Resistance*, over 10 years before, made the passage from Quebec to Cork in $12\frac{1}{2}$ days. We have little hesitation in stating that, with the same chance, the *Sovereign* would have made the same passage in 9 days—but more likely in 7; and further, with a whole-sail breeze, she would have run any ship or steamer then in the British Navy, or any other navy, out of sight in a few hours.

Donald McKay had ceased to own the *Sovereign*, so shortly after she reached Liverpool Captain Lauchlan McKay returned to this country. In about 11 months she had earned about $200,000. Captain McKay's place was taken by Captain Henry Warner, an Englishman for many years a resident of East Boston, and who had served with him as mate since his ship had been launched.

Here's a challenging advertisement which then appeared in Liverpool papers:

The clipper ship, *Sovereign of the Seas,* Capt. Warner, at Liverpool, for Melbourne. Freight, £7 per ton to the wharf, and return 50s per ton if she does not make a faster passage than any steamer on the berth here, or in London, Freight without warranty, according to agreement."

This was loud talking, when it is borne in mind that the celebrated steamer *Great Britain* was up for the same port, and had been recently refitted to render her equal to a clipper under canvas alone, and that she advertised to return 40s. per ton of freight, if her passage exceeded 65 days. Still notwithstanding all her advantages, the McKay ship beat her handsomely.

Bound for Melbourne, the *Sovereign* sailed from Liverpool, September 7, 1853, with a cargo valued at one million dollars and arrived there in 77 days, having beaten every one of the ships that sailed with her, although loaded down to 23½ feet and about one-half of the crew disabled, including the first mate.

It was stated that she carried over four tons of gold dust on a passage, from Melbourne to London, January 23, 1854. The crew arose in mutiny and made a rush aft to seize Captain Warner and his officers, and convert the ship into a pirate. Single-handed he seized a cutlass and opened a lane among them, while the three mates armed themselves, and following him, secured the ring-leaders and ironed them. The crew was mostly com-posed of Australian beach-combers, some of them ex-convicts, who were always ready to undertake anything desperate. Getting crews in Australia was exceedingly difficult in those days, when "off to the mines" rushed whole shiploads of sailors. With about half his crew in irons, Captain Warner brought his ship safely to London, where he was highly complimented by the underwriters. The *Sovereign of the Seas* and her cargo were valued at over a million dollars.

She cleared from London, August 1, 1854, for Sydney, N. S. W., under command of Captain Muller, having passed into possession of J. C. Godeffroy of Hamburg, Germany. She arrived at Sydney on October 22d, in 84 days. It was claimed that in the early part of this voyage she promised to excel all her earlier performances. On the fortieth day out she was off Cape Good Hope and but for a succession of easterly gales, would have made the passage in an unprecedented short space of time. Her best day's run was 410 miles, and her log proved her to have traveled occasionally at the rate of 22 miles an hour. On September 8th, she met with a casualty that would have resulted in a serious detention, had she not been well found and manned. She was sailing comfortably under easy canvas, when a sudden storm denuded her of topmasts and gear attached, leaving nothing but her three lower masts standing. In six days from the accident everything was repaired, and the ship on her way at her usual speed.

From Sydney the *Sovereign* proceeded to Shanghai via Hong Kong. She appears to have been in the Shanghai trade for a voyage or two, being under the German flag.

Late in 1856 she was engaged in general trading. She was reported to have been sold at London, in June, 1858, for $40,000 having shortly before undergone repairs to the amount of $12,500. She was listed as being in port at Whampoa under the British flag in December, 1858. She came to her end August 6, 1859, being wrecked on the Pyramid Shoal in the Straits of Malacca.

Thus we close the story of the *Sovereign of the Seas*. Not only did her splendid performances redound to the benefit of Donald McKay, who daringly designed, built and originally owned her, but she was the only ship that brought him, deservedly, large financial returns.

Let us before closing, review the life history of Captain Lauchlan McKay, who sailed into fame with the *Sovereign of the Seas*.

As has been told before, when Donald McKay settled in East Boston, he grouped around him his father and mother, his brothers and sisters to share his prosperity. All his brothers were skilled mechanics and men of rare energy. Lauchlan had learned his trade in New York with Isaac Webb, after which he served four years as carpenter on the U. S. Frigate *Constellation*, in which Admiral Farragut was a young lieutenant at the time. Before his appointment to the command of the *Sovereign of the Seas*, he had engaged in shipbuilding at East Boston and also sailed both as a ship's officer and commander in the trans-Atlantic trade for some years. In 1853, after relinquishing command of the *Sovereign*, he commanded the *Great Republic* on her passage from Boston to New York, where she was burnt—then the McKay brothers, Donald and Lauchlan, knew her no more.

Subsequently Capt. McKay commanded the British ship *Nagasaki* on a voyage from Liverpool to Australia. While at Sydney, N. S. W., he raised a large ship which had been sunk. Other parties had tried for a month and failed. He raised her in less than a week. On the same voyage he spoke a ship in distress, leaky, and took off all hands with their effects. He then went with his own carpenter and inspected her and by pulling around her, discovered that she had been bored through in the run.

When she pitched the splinters of the auger holes attracted his notice. It was evident that there had been foul play for the purpose of defrauding the underwriters. After considerable trouble the holes were plugged, the ship pumped out and a prize crew put on board under command of Capt. McKay's mate. She reached an East

Indian port, where an investigation took place, but could not be completed, because her original Captain, officers and crew were on board the *Nagasaki*. Capt. McKay was not only rewarded by the underwriters, but complimented by the British Board of Trade for his promptness.

Retiring from the sea, he formed a partnership with Captain Charles B. Dix, and engaged in the general shipping business at New York, later becoming extensively interested in the Greenland and Newfoundland trade, under the firm name of McKay & Dix. He retired in 1893 and removed to Roxbury, Mass., where he lived with a niece and nephew, whom he had adopted as his children, until his death, April 3, 1895. Of him it can be said, that his energy and skill as a shipmaster combined with that of his brother, Donald McKay, as designer and builder of ships, brought the two oceans more closely together perhaps than did the work of any other two individuals following like professions in the world.

"SOVEREIGN OF THE SEAS II"

A second vessel by this name, registering 1,502 tons, was built by Donald McKay in 1868, for Lawrence Giles & Company of New York. After sailing for some years under the American flag, she was sold to German owners and renamed *Elvira*. Again known by the name she first bore, the *Sovereign II* was purchased by Lewis Luckenbach of New York and converted into a coal barge about 1903. She travelled up and down the Atlantic coast, causing more or less discussion as to her identity, many persons contending she was the famous McKay clipper built in 1852. In 1910 she was lost off Barnegat in a gale of wind.

CHAPTER XXIII

IF the Irish famine demonstrated what Boston was capable of doing in an emergency, certainly the discovery of gold in California afforded a better test. The first to avail themselves of the commercial opportunities offered, were the people of New England; in the new settlements, which were being peopled rapidly, the materials required for necessity and comfort were first demanded; luxuries being a secondary consideration. Enterprising Yankee merchants comprehended the fact, and also the advantage of first entering the field, which they did with great energy. Boston ships began early to play an important part in reducing the length and dangers of a voyage around Cape Horn, which resulted in inaugurating the fleet of clipper ships, destined afterward to exert so great an influence both in American and Foreign commerce. The California trade indirectly benefited the East India and China trade of Boston and New York as there were no return cargoes from the Pacific, and the ships sought the East Indies and China for a homeward freight.

No sooner had the *Flying Fish* proven herself than her owners commissioned Donald McKay to build another clipper ship for their Californian trade. To commemorate England's naval and commercial glory when her enterprise was spreading under Queen Elizabeth's reign, Charles Kingsley wrote *Westward Ho!* and now Sampson & Tappan selected that appropriate name for McKay's latest creation.

With very rakish masts, three skysail yards and carrying royal studding sails, this magnificent ship certainly must have presented a beautiful sight as she swept past Boston Light bound for San Francisco. Her length on the keel was 194 feet; on deck, between perpendiculars, 210; and over all, from the knight-heads to taffrail, 220 feet; extreme breadth of beam 40½ feet; depth of hold 23½, including 7 feet 8 inches height of between-decks; deadrise at half floor 20 inches, swell 6, and sheer 2 feet 20 inches. Her ends were long and very sharp, with slightly concave lines. There was not a straight place in her model. She was planked smoothly, without a projecting line, up to the covering board, and had neither head nor trail boards. The full figure of an Indian warrior, represented as advancing rapidly in the chase, placed on a pedestal of ornamental flowering, with her name on each side of the bow, were her principal

Boston May 6. 1856.

Donald McKay Esq,

Present,

dear Sir,

Understanding
that you propose Embarking tomorrow in
the Steamer for England, we take great
pleasure in giving you this line to say that
we have had built under contract with
you the Clipper Ships Stag Hound, Flying Fish and Westward Ho' amounting to
four or five thousand tons - that the contracts
on your part were most satisfactorily carried
out, that the Ships have performed Equal
to any ever built, have always discharged
their Cargoes in fine order and reflect
great Credit on your Skill and
thoroughness as a ship builders - Were
we to build tomorrow there is no man

"WESTWARD HO"

Facsimile.   Letter from Sampson & Tappan of Boston recommending Mr. McKay, etc.

in the New England States we should do
from contract with as yourself, feeling
assured that we should get a superior
ship, be most honorably dealt by & have
on your part a liberal fulfilment of
the Contract —

Wishing you a pleasant trip across
& safe return, We are Dear Sir

Yours most truly

Sampson & Tappan.

"WESTWARD HO"

Facsimile. Letter from Sampson & Tappan of Boston recommending Mr. McKay, etc.

ornaments forward. Viewed either end or broadside on, it seemed almost impossible to conceive of a more perfectly beautiful model. Nothing which skill and money could command to render her strong and durable, had been withheld. A more beautiful or better built ship of her size could not be desired.

In this connection we reproduce fac-simile copy of letter given to Mr. McKay, by the owners of *Westward Ho!* a few years afterward.

Captain Edward Nickels of the *Flying Fish*, in which he had just previously arrived from the Far East, was to take command of the *Westward Ho!* He was justly esteemed as one of the best sailors afloat, but Captain Johnson was placed in command of what was considered the crack Boston ship up for the port of San Francisco. It was confidently expected that her maiden voyage would be made under 100 days.

All this while a rivalry was going on in the California trade. No vessels cleared together or within a fortnight of each other from the ports of New York and Boston for the Golden Gate that were not the subject of racing wagers.[1] Sailing from Boston on October 20th, 1852, the *Ho!* had the long passage of 29 days to the line. Passing Cape St. Roque 31½ days out, she had the good

---

[1] It followed in the natural order of events therefore, that the *Westward Ho!* was a contestant in that famous race around the Horn against the clippers *Flying Fish* and *John Gilpin*, *Wild Pigeon* and *Trade Wind*, leaving New York at approximately the same time.

run of 21 days to 50° South and 13 days afterwards found her bound north in the Pacific; 24 days later she was on the equator, 88 days out, and arrived in San Francisco harbor, February 1st, 1853,—making 103 days out. The *Flying Fish* won the race in 92 days, 4 hours from port to anchor and Captain Nickels must have derived much satisfaction in outdistancing so greatly the new crack Boston ship, which was to have been placed under his command.

At the time of which we write there were many large square-rigged vessels lying in San Francisco bay, unable to leave for want of hands. With sailing ships of all nations entering the Golden Gate in a steady procession, AB's were as scarce as bananas at Point Barrow and skippers were willing to pay most any price to get the required number of certified hands. Hence many a "lemon pelter" and "square head," just arrived after a long passage from "The Lizard", say, and desiring to remain ashore a couple of weeks at least to wet his whistle before shipping again, found himself involuntarily bound out once more on another ship for China, within a few hours after he had signed off and had tipped his first horn with the smiling "shipmate" with the tall hat and gold-headed cane containing the old knockout drops.

Many of the vessels that arrived at San Francisco during the mad rush for gold, never got away, but in a few years rotted and tumbled to pieces where they were moored. As stores and dwelling houses were much needed, a considerable number of deserted ships were drawn high on the beach, and fast imbedded in deep mud, where they were converted into warehouses and lodgings for the wants of the crowded population.

From San Francisco the *Westward Ho!* sailed to Manila in 39 days, beating *Flying Fish* one day on the run. She continued on to Batavia and was thence 82 days to New York; had touched at St. Helena and was 35 days from there to port. In 10 months, 8 days she had sailed around the world. Making allowances for slow time between Manila and Batavia because she faced a contrary monsoon, and including about two months detention at the three ports visited, this must be considered a very good voyage.

On her second voyage the *Westward Ho!* sailed from New York, under command of Captain Hussey, November 14th, 1853, arriving at San Francisco, February 28th, making the passage in 106 days. A slow run of 70 days to Singapore followed. From Calcutta she went to Boston in 103 days, thus again circumnavigating the globe in one year and twelve days.

On April 24th, 1855, the *Ho!* arrived at San Francisco from Boston after being out 100 days and 18 hours. The clipper *Neptune's Car*, Captain Joshua A. Patten, passed out of Sandy Hook the day after her competitor left Boston and much interest was taken by shipping men in the outcome of this voyage. Her passage was made in 100 days, 23½ hours. Afterwards there followed long and spirited arguments as to which ship had the advantage, some authorities claiming that a departure from Boston was a disadvantage as against a sailing from New York, while others expressed the opposite opinion. It was never satisfactorily decided. Both vessels, however, proceeded to Hong Kong, and in what may be classed as a drifting match more than a test of sailing powers the *Neptune's Car* beat the McKay flyer 11 days on this run across the Pacific, being 50 days against 61. An anticipated continuation of the race from China to England was prevented by the *Westward Ho!* taking a cargo of living freight—Chinese coolies from Swatow to the Chincha Islands guano deposits.

American clippers were now engaged to a large extent in this—a traffic worse than the African slave trade. The British Government forbade it from a consular port—but Swatow was out of its jurisdiction, and through the active co-operation of the Peruvian consul, great numbers of these poor wretches were taken away and remained in the iron bondage of Cuba or South America. We note the *Westward Ho!* having passed Anjer with 800 coolies en route to Callao; others passing continually—English coal ships, Peruvian convict hulks, and American clipper ships, all headed towards the west coast of South America, every square foot of space occupied by a poor devil of

Westward Ho! in the China Sea. From painting by Charles Robert Patterson. Reproduced through courtesy of George W. Rogers, Esq.

a Chinaman, who thought, when he received a dollar in hand to be spent in clothing, and made a contract to work five years at $8 per month—by paying $50 for a passage, with all the rice he wanted guaranteed—that he was leaving purgatory for paradise! But when his owners put him to work on the guano deposits, under the burning sun of the Chinchas, he found out how sadly he had been deceived!

The *Westward Ho!* arrived at Callao February 4th, 1856. Captain Hussey did not appear to have had any trouble with his human cargo, and after discharging them, his ship was loaded with guano and took her departure from Callao on June 12th for New York.

Another very fast passage of 100 days was accomplished by the *Westward Ho!* on her fourth and last voyage from New York to San Francisco, where she arrived March 26th, 1857. At the last named port, the command was transferred to Captain Jones, who took her down to Callao and she continued as a coolie slaver. Having been purchased by Peruvians she went under their flag, hailing from Callao, and her name, an unusual thing for foreigners, remained unchanged. Most of her latter day operations are understood to have been between China and Peru. Her end came when she caught fire in the harbor of Callao and burned until she sank at her moorings, proving a total loss.

CHAPTER XXIV

I F the reader had been at East Boston on a certain
day in November 1852, he would have found this
beautiful clipper on the stocks in McKay's yard,
ready to be consigned to her watery element.

In spite of its technique, so charming a description of
a "deep sea lady" was undoubtedly considered a classic
in the sailing ship era, when that colorful descriptive
writer, Duncan McLean, contributed this to the Boston
Atlas:—

THE NEW CLIPPER SHIP *BALD EAGLE* OF BOSTON

On the keel she is 195 feet long, between perpendiculars on
deck 215, and over all 225; her extreme breadth of beam is
41½ feet, and depth 22½, including 8 feet height of between
decks. In model she differs widely from any clipper which we
have inspected. The rise and form of her floor are designed
to obtain the greatest possible buoyancy consistent with stability and weatherly qualities. Her lines, too, have been formed
upon the principle that when sailing by the wind, the pressure
aloft will incline her, and to overcome the consequent angular
resistance, is one of the elements in her model. But whether

sailing, inclined to the plane of the horizon, or at right angles to it, her lines have been calculated for both, so that she is expected to float more buoyantly and pass more easily through the water than any other clipper that has yet been built. At the load displacement line, she is sharper than any other clipper, and her lines, for twenty feet from the cutwater, are almost straight, but aft they swell into the convex, and blend beautifully with her fullness amidships. Her greatest breadth of beam is at the centre of her loadline, and her lines aft are decidedly convex. She is fuller aft than forward, upon the principle that, when passing rapidly through the water, she will be liable to settle aft, hence the fullness of the lines to buoy her up; and also, that the pressure of the water, as it closes aft, will actually force her ahead, and leave her without a ripple. Her model above is also designed with special reference to overcoming atmospheric pressure; hence she has little if any flare to the bow, (which is angular in its outline to the rail), low bulwarks, and a flush deck. Her bow is long, very sharp, and rises grandly in its sheer; and the cutwater is just inclined enough to make her a perfect picture forward. She has a large gilded eagle on the wing, for a head, and it forms the best and most beautiful head that we have yet seen upon any clipper. The ends of her catheads are ornamented with gilded carved work; otherwise she is smack-smooth forward.

She has about three feet sheer, and sufficient swell or rounding of sides, to preserve the harmony of her lines, and she rises forward and aft with such easy grace, that even on the line of the planksheer, the eye cannot detect any wavering in its sweep. Her stern is slightly elliptical, inclined aft, and is formed from the line of the planksheer, the moulding of which and the strake below, form its base. It is very light, beautiful in outline, and tastefully ornamented.

The *Bald Eagle* was owned by George B. Upton of Boston and placed under command of Captain Philip

Dumaresq, appropriately called—the "Prince of Sea Captains." He was a man of wealth and related to some of the most prominent families in Boston, following the sea because he had a positive fondness for seafaring—an ocean voyage to China in his youth, under the advice of a physician, had been the means of restoring him to health, —and at 22 he took command of a vessel. He was not only a thorough sailor, but one of the most accomplished navigators in the world. Captain Dumaresq relinquished command of the *Surprise* to take the new and larger *Bald Eagle*.

Every new experiment in the maritime world was now keenly watched by a multitude of eager men. Month after month, ships surpassing in beauty and strength, all that the world had ever produced, were built and equipped by private enterprise, to form the means of communication with California's new land of promise. Our eminent shipbuilders and most enterprising merchants vied with one another to lead in the great race around the Horn. Established rules which had for years circumscribed mechanical skill to a certain class of models, were abandoned, and the capitalist only contracted for speed and strength. Ships varying in size from 1500 to 2500 tons were soon built and sent to sea, and their wonderful performances, instead of satisfying increased the desire to excel.

So many beautiful clipper ships could be found in the early fifties amongst the shipping which lined the docks of New

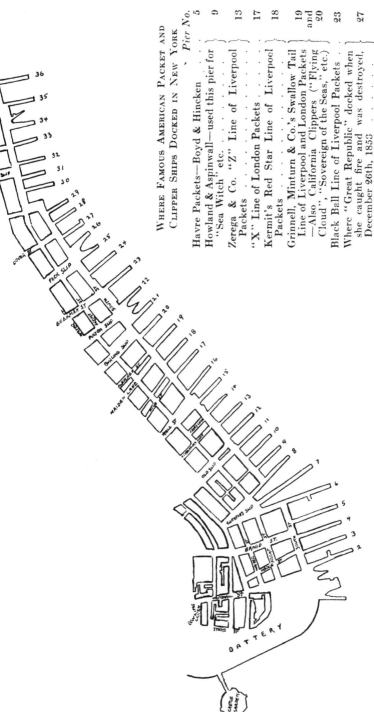

WHERE FAMOUS AMERICAN PACKET AND
CLIPPER SHIPS DOCKED IN NEW YORK

	Pier No.
Havre Packets—Boyd & Hincken	5
Howland & Aspinwall—used this pier for "Sea Witch", etc.	9
Zerega & Co. "Z" Line of Liverpool Packets	13
"X" Line of London Packets	17
Kermit's Red Star Line of Liverpool Packets	18
Grinnell, Minturn & Co.'s Swallow Tail Line of Liverpool and London Packets —Also California Clippers ("Flying Cloud", "Sovereign of the Seas," etc.)	19 and 20
Black Ball Line of Liverpool Packets	23
Where "Great Republic" docked when she caught fire and was destroyed, December 26th, 1853	27

MAP OF EAST RIVER WATERFRONT

South Street from Battery to James Slip showing shipping district, piers, etc., in 1852. Taken from photographs in the McKay Collection. Original maps owned by New York Historical Society.

York's East River waterfront, that we will take a glance at some of the points of contrast between them and the old-fashioned clumsy craft that they superseded.

Beginning forward, as is a sailor's bounden duty, let us take a look at the head gear. In the first place, the modern bowsprit steeves very much less than the old one. By this means the leverage of the head sails is greatly increased, and the ship looks better. It is not quite so comfortable, however, in reefing or furling the head sails. Next, the old ship always carried a heavy, awkward spritsail-yard, of no earthly use but to strike porpoises from, and to hang clothes upon to dry. This is now entirely dispensed with. A great, two-legged martingale also encumbered the head gear, while now a small delicate, single one is the style. The bobstays, bowsprit, shrouds, and other rigging in that vicinity were of heavy rope; now they are of chain or iron rods, exquisitely neat and of ample strength, and when properly set up require no lifting for years.

Next noteworthy is the ground tackling. Instead of the neat chain now shackled to the ring of an iron-stocked anchor, there was a hard, harsh, hempen cable, bent on with all sorts of seizings to a great wooden stocked anchor. It was no child's play, in old times, to handle a ship's ground tackling. To "overhaul a range of cable" in cold weather, was a forenoon's job; now it is done in a few minutes. To heave it in of a cold day; and stow it away in the cable tier, was positively the very worst work a crew could do. Now, a few minutes at the new-fashioned windlass and the anchor is apeak, and the cable let run into the chain boxes without delay. The old-fashioned ground tackling was not a great deal better in hot weather, though one's fingers did not get quite so benumbed in handling it. But even then it was a hard dispensation, for the worms and barnacles of the tropics and other warm latitudes would form colonies on the hemp cable; and it was Jack's Sunday privilege to give it a scrubbing, by just dropping another anchor under foot and heaving in the old one. Jack's theology obliged him to make an addition to a certain precept of the Decalogue, that read in this wise—"Six days shalt thou work and do all you are able, on the seventh heave up and scrub the cable."

In old times ships had no bulwarks forward of the fore rigging. It was all open there, why, heaven only knows. Now the bulwarks are much higher than the tallest man's head, and solid all around forming a capital breakwater.

A few feet further aft we come to the windlass, and here is a wonderful improvement. The old windlass was merely a large round piece of wood, with holes in it for handspikes, and a few iron pauls. It had no leverage power except that derived from the arm of the handspike. At this miserable machine the sailors had to heave and heave until they were exhausted, and even then they were frequently unable to lift the anchor. The modern windlass is an instrument of great power, with brakes like those of a fire engine, by a few strokes of which the most firmly embedded anchor is lifted from its oosy bed.

Still passing towards the stern of the ship, the long-boat meets the eye. Formerly it was stowed directly over the main hatchway, requiring it to be removed before the cargo could be got at. Now it sits in its chocks from one voyage end to another, well housed over.

The *Bald Eagle* had a small top-gallant forecastle; and abaft the foremast a house 36 feet long by 8 wide. Its forward division contained a metallic life boat, placed on rollers, and the sides of the house were made to unship, so that the boat could be taken out upon either side of the deck, when required.

The galley too, is a different affair from that of olden times. It is more like a well-appointed kitchen than a galley. Many is the gale of wind in which the "doctor" (as the cook, even in those degenerate times was called) could not boil water enough in the old galley to mix with grog, what the crew felt in duty bound to drink to "Sweethearts and Wives" on Saturday nights. When the fires are put out in the galley of a modern built ship, it will be because she has taken a sudden freak to pay Davy Jones a visit in that unexplored locker of his.

On the quarter-deck the most important improvement is the steering apparatus; and very great is the debt due to mechanics on shore for the invention and perfection of the new wheel; it is now easy work to steer the largest ship, even by one man, and with accuracy too; but in old times it took two men when scudding, to keep the ship within an average of three or four points of her course. The rudder is now fastened to the stern-post with great strength and neatness, and cannot by any possibility get unshipped by accident.

Now let us cast our eyes aloft. Those iron trusses first attract attention. How freely the lower yards swing upon them; requiring no overhauling for stays, as did the old, greasy, creaking rope truss. Chain has taken the place of rope for topsail ties, topsail and top-gallant sheets, and iron rods for futtock shrouds. What with the large use of iron where rope was formerly used, and the superior manner in which shore riggers do their work, there is very much less work to be done on the rigging while at sea than of old. A ship is also handled easier in consequence.

It is curious to observe the difference in the management of matters on the day of sailing now as compared with olden times. Then the vessel hauled into "the stream," and anchored there, to remain a day or two, to see that everything was all right, that nothing was forgotten; as well as to get the crew sober. When the captain found everything right, the crew tolerably sober, and the wind fair and the weather settled, he would loose his foretopsail as a signal to the pilot that he was ready to go to sea. Whereupon a staunch, broad-beam and weather-beaten pilot would come on board and take the ship out. Now a steam tug fastens itself to the ship's side, and she goes to sea from her berth at the wharf. It would be losing too much time to sober the crew at anchor.

At New York the *Bald Eagle* loaded in Ogden's line of San Francisco clippers, and sailed from that port December 26th, 1852, and considering the winds she encountered, made a remarkably short passage to California. She beat every vessel which sailed about the same time. This undeniably was accomplished in great part through the indomitable energy of her commander. Captain Dumaresq never undressed but to change his clothes, nor slept in a bed, the whole passage. Like a horse he often slept standing up, with an eye and an ear open. Some idea of his labor may be formed from the fact that, though enjoying perfect health, he lost 34 pounds weight during the passage of 107 days. His log was perfect to a fraction, and the following facts taken therefrom, convey some idea of the *Bald Eagle's* maiden voyage.

She was 29 days to the Equator, and crossed it in longitude 31° 13' and her best day's work to that point was only 230 miles, and only during 4 days did she average 220 miles, the rest of her days' works are very small. On the 16th of January, Capt. Dumaresq writes: "Seventeen days head winds, tacking as the wind hauled." From the line to 30 S. lon. 46 36 W. there are only two days over 200 miles and most of them less, the average being for 13 days only 172 miles. February 15th, 52 days out, passed the Straits of Le Maire, having averaged to that point only 154 miles per day. March 15th, lat. 23 44 S. lon.1 01 41 W., Capt. D. remarks: "For 22 days winds at N.W. almost dead ahead:" and again on the 21st of March says: "Got the S.E. trades in 20 S. and lost them in 5 S. On the 23d March crossed the Equator in lon. 111 15 W—88 days 3 hours out. April 11th, 1 P.M. made the Faralones: at 5:30 took a pilot, and at 9 P.M. passed the Heads, making 107 days and 9 hours passage." Capt. D. sums up by the following: "Distance sailed, by observation, 15,802 miles, averaging about 147 miles. Only a portion of 43 days with the wind fair enough to set studding sails, and for 22 days averaged only 66 miles. The ship has never been under close reefs, nor have the courses, spencers, or spanker been furled during the passage."

Captain White of the ship *Flying Childers*, sailing from Boston a week before the *Bald Eagle* left New York, was more fortunate, yet she beat the *Flying Childers* six days, and both beat the fine New York built clipper *Jacob Bell*.

Sailing from San Francisco May 8th, the *Bald Eagle* arrived at New York August 13, 1853, 96 days passage. Captain Dumaresq left her at New York to take command of the new clipper *Romance of the Seas*—the sharpest ship ever built by Donald McKay or any other shipbuilder, in our opinion.

Captain Caldwell took over the *Bald Eagle*, had a run out to San Francisco of 115 days, arriving January 25, 1854.

San Francisco was passing through a period of much mercantile distress. For some time commercial ventures had been unprofitable. An excessive quantity of goods had arrived during the latter part of 1853, and importations early in 1854, continuing very large, the market was completely glutted, and prices of the great staples of commerce fell day by day. In consequence the *Bald Eagle* and some other vessels were fully freighted to New York and other eastern ports with goods similar to those they had recently brought from thence. Sailing through the Golden Gate on March 1, 1854, she arrived in New York harbor May 19th, 79 days afterward, within 3 days of the record, which was made by a ship in ballast.

The *Bald Eagle* sailed to San Francisco again in 115 days. She then went to China in 47 days, under Captain Treadwell.

It was at this juncture that she was engaged in the Coolie trade. Like many other ships in the Californian trade, she went in ballast from San Francisco to the Chincha Islands for guano, where many homeward bound Australian traders also picked up their cargo. But it was given greatly to certain Boston clipper ships, owned and managed by New Englanders who at home professed to be Christians, to materially assist in this devilish business.

Whittier by the following stanza in his poem, "The Shipbuilder," indicates a knowledge of this Chinese Coolie traffic, as well as the loathesome opium trade:

Speed on the Ship! but let her bear,
No merchandise of sin,

No groaning cargo of despair,
    Her roomy hold within.
No Lethean drug for Eastern lands,
    No poison draught for ours;
But honest fruits of toiling hands
    And nature's sun and showers.

Here is an account by Captain Hayes of the ship *Otranta* trading between San Francisco and Singapore, etc. Referring to the examination of the poor coolies on board the junks that took them off to the ship as too revolting for description, he continues:—

All men over thirty-five years old, or after they have been stripped stark naked show the least sign of disease upon their persons are rejected, and these poor creatures, brought a long way from the interior by "crimps" of their own nation— who get $10 for bringing down all of what they term healthy cattle—are turned ashore to perish of starvation or die a lingering death by exposure. Great numbers, says Captain Hayes, are seen along the beach in this horrible state. Perhaps, he added, they are far better off than those poor wretches who have been led to suppose they are bound to the golden regions of California or Australia, or some pleasant island in the China or Indian seas. The moment they are passed and get on board the ship, they have the sulks and want to go back; but no, they had crossed the Rubicon, and must remain in the iron bondage of Cuba or South America. When mutiny is among them, the Captain credits only the interpreter, or the one who makes the fact known. This man, therefore, has the power to so misrepresent the feeling on board as to occasion strict and harsh measures, against which they rise. The most danger arises before they pass the land; afterwards, the boundless look of ocean and their respect for navigation, under kind treatment, will usually keep them in their place.

It is said that the *Bald Eagle* left Swatow about the same time as the *Westward Ho!*—each ship with about 700 Chinese coolies aboard; the former bound for Havana, the latter to Callao. Captain Treadwell duly delivered his cargo of human beings.

There is a fake yarn about the destruction of the *Bald Eagle*, given as authentic by a British writer with such luridness, that we publish it. Reciting the truth about a ship's history becomes prosaic at times.

This is proved by the terrible tragedy of the clipper ship *Bald Eagle*. Like many another fine ship she gravitated into the coolie trade, and not the highest but the lowest form of coolie trade—that of carrying the refuse of China to that hell whence they never returned, the Chincha Islands. She was, in fact, but little better than a slaver. For years she drudged steadily at this awful trade, sinking lower and lower in the social scale of ships until at last a time came when even her officers were foreign, and the only sign of her past glories was the star-spangled banner which still flew from her monkey gaff.

On her last and fatal voyage her captain was a Portuguese, and he likewise shipped a crew of dagos, mostly his own countrymen, the only Northerner being an Irishman, who was responsible for the terrible account of her end. How much of his yarn was an exaggeration it is impossible to say, but, knowing of one or two other not dissimilar tragedies on coolie ships of that time, I should say very little.

The *Bald Eagle* was 500 miles east of Manila, bound for Callao, and reeling off an easy 10 knots under the influence of a stiff breeze. It was five bells in the afternoon watch, and all seemed quiet below, when suddenly a wild screech rang out, and the next moment an avalanche of Chinamen attempted to rush the hatchway ladders, having torn down their bunk boards for weapons. The crew, however, were just in time to keep the maddened Celestials off the deck by fastening down the hatch gratings.

Then the captain, being a Portuguese, acted as such, and bringing out his revolvers began shooting through the gratings at the wretched coolies, the mates following his example. But even shooting rats in a trap is sometimes dangerous, and so it proved on this occasion. The Chinese were in such a frenzy that they cared nought for the bullets, and stood out under the hatchway grating, cursing and shrieking at the shooters until there was a wriggling mass of dead and wounded Celestials piled up almost as high as the iron bars. And this was the cause of the final tragedy. So close were the pistols to the pile of dead Chinamen that a spit of flame actually set a light to the clothing of the uppermost. Immediately there was a furious rush to obtain the burning cloth, and the maddened coolies fell over one another, entirely heedless of the bullets, in their eagerness to preserve the smouldering piece. It was soon torn from the dead man's shoulder, the man who got it at once blowing upon it to keep it alight. A bullet stopped his efforts, but another seized it only to be shot in his turn; and so the murderous business went on with the cloth still alight. As fast as those above shot down the men who held this fatal fuse others filled their places, until at last the tiny flame, which had been kept alight at such a cost, disappeared from beneath the hatch, still burning.

Half an hour later smoke began to ascend out of the fore and main hatches. The crazy Chinamen had set the ship on fire, evidently thinking that this would compel the crew to take the hatches off and thus give them their chance to rush the ship and capture her. But the Portuguese had no intention of taking any such risk. Instead they cut small holes through the deck, and began to pump water below with the aid of the wash-deck hose. With hundreds of infuriated coolies intent on keeping the flames alight, this was, of course, a useless proceeding, and in a very short while the fire had so increased that the heat and smoke compelled the Chinese to crowd under the hatchway gratings. But when they found that the crew had no intention of letting them up on deck their despair may be imagined, as it had become too late for them to be able to put out the flames themselves.

The scene now grew worthy of Dante's Inferno. Beneath the bars the wretched Chinese struggled in a seething, wriggling mass of terrified humanity, packed as tight as sardines by their desperate mates further back in the heat and smoke. From this mass a long-drawn shriek of terror rang shrill and piercing into the growing darkness. To those who looked from above nought could be seen but a sea of faces turned a sickly green with fright, their eyes starting out of their heads, and their mouths opened wide as they gave vent to one horrible endless yell. As the flames approached closer and closer to the hatchways, another frightful element was added to the tragedy, and that was the awful smell of burning flesh as those on the outskirts of this human maelstrom under the square of each hatch succumbed to the fire.

The crew had long since ceased to pump water, and were now only concerned in getting safely clear of the ship. The *Bald Eagle* was hove to just as night fell with great difficulty, for the smoke pouring out of the deck was so dense that the men could scarcely breathe and had to work as if in a thick fog, at the same time the deafening shrieks made it impossible to hear the orders of the officers.

By 8 o'clock the *Bald Eagle* was in a blaze fore and aft in spite of torrents of rain, which had begun to fall at sunset. Slowly the yells of the burning Chinamen had died down until a ghastly silence reigned, the last coolie having succumbed in the fiery furnace below the grim bars of the hatch gratings.

With furious haste the crew now set about launching the boats, into which they only had time to place a little biscuit and water, barely enough for one square meal. One of the boats was stove in being got over the side, so that when they at length pulled away from this awful crematorium the two quarter boats had ten men apiece, and the gunwales of the long boat were almost awash with eighteen men. The long boat had masts and sails, but the quarter boats only oars, so it was decided to tow them. The captain shaped a course for Manila. The wind was dead aft, fresh, and with a heavy following sea. All that night the long boat ran before it with the other boats in tow, all three having many narrow escapes from capsizing or being swamped.

Three nights and two days were passed in this fashion, with only the nibble of a biscuit for each man and the scantiest supply of water. And, as if this was not enough, the superstitious Portuguese were terrified by the continual presence of a large shovel-nose shark, which kept pace with the long boat, now on one side, now on the other.

On the last night the tow line of one of the quarter boats parted, and she was afterwards found stove in and floating bottom up, though there were no signs of her crew, who, it was surmised, were eaten by the shark.

Early on the morning of the third day the land was made ahead, only to be blotted out the next moment by a dense mist. However, now, for the first time, the wind dropped and fell light, and the two remaining boats presently found themselves entering the harbour of Manila. Here they found H.M. gunboat *Rattlesnake*, which took them on board and looked after their wants. Such was the end of one of the most horrible tales of the sea it would be possible to imagine.

For the following detailed history of the *Bald Eagle*, we are indebted to that well-known authority on ships and shipping of the past, Mr. F. C. Matthews of San Francisco.

On her fourth voyage the *Bald Eagle* sailed from New York July 18, 1856, and arrived at San Francisco after a passage of 120 days, beating the extreme clipper *Neptune's Car* 16 days. Captain Treadwell reported having had light or baffling winds most of the passage; off Cape Horn, however, heavy weather was experienced; the run up the Pacific occupied 53 days. From San Francisco the run to Calcutta was made in the excellent time of 59 days; from Calcutta to Boston she was 98 days.

In 1858 and 1859 the *Bald Eagle* was not engaged in the California trade, having been diverted to the China run. She went out from Boston in 108 days, and after trading coastwise for about a year loaded for Liverpool, reaching that port on December 21, 1859, after a passage of 120 days from Hongkong. Under date of March, 1860, it was stated that she was expected to arrive at Foochow under command of Captain Nickels, formerly of the *Flying Fish*, about June 1 under charter to the British government, with coals and stores for their naval forces, which she had loaded at Liverpool. She reported at Shanghai on June 25, having had the long passage out to Anjer of 94 days; her time thence to port of destination, 32 days, was also very slow, if she went direct. She remained on the China coast during the fall and winter of 1860. We find that she arrived at San Francisco on April 25, 1861, 41 days from Hong Kong, and this was her last visit to that port. She sailed on June 16, touched at Honolulu on the 13th day out and thence had a long and tedious passage across the Pacific of 56 days, encountering nothing but light winds and calms throughout.

Captain Nickels relinquished command at Hong Kong and later died there of yellow fever. Under Captain Morris, the *Bald Eagle* sailed from that port on October 15, 1861, for

San Francisco and has never since been heard of. Several very severe typhoons were experienced in the China Seas between her sailing date and the end of December and it is supposed that she foundered with all hands. She had a valuable cargo, including about 600 tons of rice, 500 tons of sugar, some tea, etc., besides $100,000 in treasure. She was still owned by George B. Upton and was in good condition, being rated A 1½. The insurance paid on vessel and cargo was $300,000. A report was published in Eastern newspapers to the effect that a portion of the crew had been picked up by Japanese fishermen, but these statements were proven to be a baseless rumor.

CHAPTER XXV

THE *Empress of the Sea* was an aspiring name, indeed, to be given in those days of shipbuilding competition, but this Boston-built clipper, seemed worthy of it. Originally, Donald McKay designed to build and sail her on his own account, therefore he modelled her according to his own ideas, and embodied every element of perfection his extensive experience could suggest. While still on the stocks, however, he sold this magnificent ship to Messrs. William Wilson & Sons, eminent merchants of Baltimore.

Notwithstanding her vast size, she looked as light and graceful as a yacht. She had three decks, was 230 feet long on deck, and 240 feet over all, from knightheads to taffrail; had 43 breadth of beam, and 27 feet depth. Her run was long and clean, formed with special reference to buoyancy, so that she should not settle aft, however fast she might fly through the water. She had 3 feet sheer, a foot curvature of sides, and a long and buoyant floor, with an angle of 23 degrees deadrise. She was

"EMPRESS OF THE SEA," 2200 TONS

Built in 1853.   From a picture in the McKay Collection.

strongly built of the very best seasoned white oak and hard pine, fastened and finished in the best style of workmanship.

In the beautiful proportions of her spars, their strength and the perfect style of their rig, with sails to correspond and finished without regard to cost, she was unquestionably one of the best fitted merchant ships aloft in the world. In all her outfits she was most liberally found.

The *Empress of the Sea* proceeded to New York and there loaded in J. S. Oakford's line of San Francisco clippers. Leaving New York, March 13, 1853, under command of Captain M. E. Putnam, she reached her destined port in 121 days, although made crank by a heavy deckload. Independent of houses, water-casks, etc., she had 100 tons of boilers upon her upper deck, and was consequently so crank that, when the wind was on the beam, she was compelled to double-reef her topsails, when she ought, if not overladen, to have carried all sail. Notwithstanding this, her passage speaks highly for the skill of her designer.

The *Empress* went from San Francisco to Callao, thence to New York, where Captain Oakford was given command, taking his ship to Quebec, then London. A voyage of 97 days to Bombay followed and returning to England, she sailed thence to New York.

Then came another voyage around the Horn, and she arrived at San Francisco June 3, 1856,—115 days. This is her quickest California passage, for her last was made in 124 days. She only made two or three more voyages under the American flag.

Sailing in Baines' Australian Black Ball Line, this ship, on June 1, 1861, left Liverpool and arrived at Melbourne August 6, 66½ days out. She was burnt at Queenscliff, December 19, 1861.

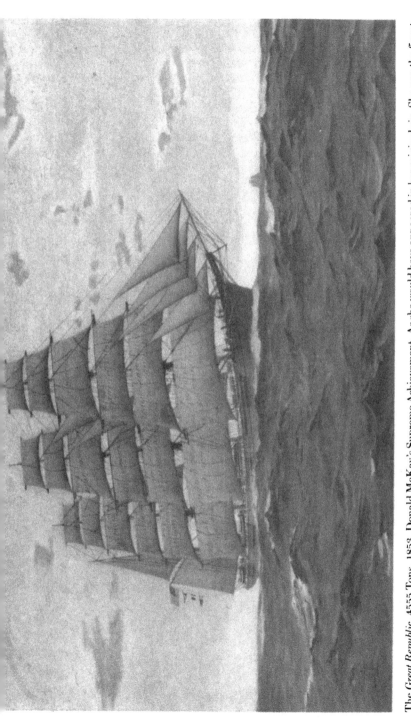

The *Great Republic*, 4555 Tons, 1853, Donald McKay's Supreme Achievement. As she would have appeared in her original rig. She was the finest, largest—perhaps the swiftest sailing vessel in the world, and in an era of ships she indeed symbolized the growing strength and greatness of our Country. From painting by Charles Robert Patterson. Courtesy of W. R. Laidlaw, Esq.

CHAPTER XXVI

"GREAT REPUBLIC," 4555 TONS REGISTER, 6000 TONS
STOWAGE CAPACITY
LAUNCHED OCTOBER 4, 1853

But deep water was never to behold the *Great Republic* just as
she sprang from her master builder's brain.

WM. BROWN MELONEY
in the *Heritage of Tyre*.

TO triumph over them all—to excel all his other
works, was the object Donald McKay had in
view when he designed this, the largest merchant ship in the world.

It was a matter of just pride with him that he had,
from his own resources, constructed a vessel, which for
size, beauty and strength, excelled the finest specimens of
Marine Architecture which ever floated upon the waters
of the ocean. In an era of ships she indeed symbolized
the growing strength and greatness of our country. She
furnished tangible evidence of America's resources and
the enterprise of American mechanics. Her strong timbers told of our forests, and her fine model and workman-

ship spoke loudly in praise of the genius and skill of our shipbuilders.[1]

When her keel was laid, some of his friends remarked that she was too large, and that she would bankrupt him before she was finished; yet he persevered, remarking by way of reply,—"*Let friends and foes talk, I'll work.*" He not only finished the splendid ship, but had her loaded, ready for sea, without borrowing a dollar from anyone upon bottomry. The *Great Republic* was clear of all encumbrances and Donald McKay owned every timber-head in her, when ready to embark upon her maiden voyage.

It was McKay's secret desire to launch the *Great Republic* upon his forty-third birthday, September 4, 1853, but unforeseen circumstances prevented. There was a sudden scarcity of proper timber and an artificial boosting of prices therefor, locally, so that he had to send far away for material, and it was exactly a month afterward before his mammoth ship slid down the ways. Her launch was an event of no mean importance in the annals of our own country's progress. In Europe, where the

---

[1] No doubt Longfellow's poem "The Building of the Ship" had something to do with Donald McKay naming his supreme achievement—*Great Republic*. Long before he contemplated building her, he had heard the famous actress, Mrs. Fanny Kemble, read this poem to a large, enthusiastic audience in Boston. Standing out upon the platform, book in hand, trembling, palpitating and giving every word its true weight and emphasis, her recital made a vivid impression upon him. It must be conceded that the launch and the construction of his masterpiece of marine architecture were virtually an embodiment of the beautiful sentiments expressed in Longfellow's verses.

LAUNCH OF THE "GREAT REPUBLIC" AT DONALD McKAY'S SHIPYARD

Boston made a public holiday of this launching, October 4th, 1853.

From an old print in the author's possession.

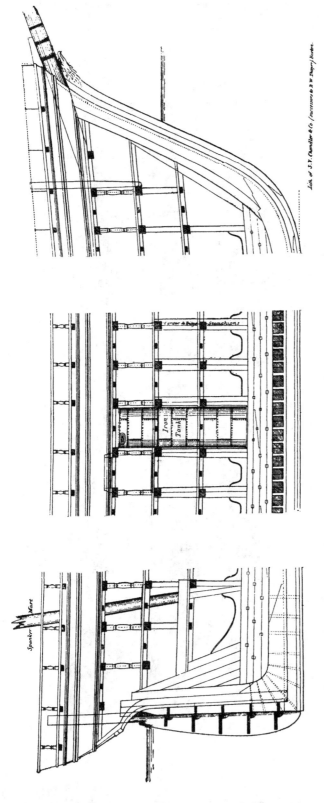

THE "GREAT REPUBLIC" 4,555 TONS—AS ORIGINALLY DESIGNED AND BUILT. DRAWING SHOWS CONSTRUCTION IN SMALLEST POSSIBLE SPACE

The above plates represent a fore and aft vertical view of the ship amidships, showing side-views of the keel, mouldings of the floor timbers, depths of the midship keelsons, stanchions and their knees, beams, ledges and carlines, outlines of the decks and rail, stem, sternpost, and rudder and positions of the spanker mast and tanks.

Reproduced from plans in the author's possession.

fame of this ship o' ships had already travelled, her coming was anxiously looked for, especially in the Mersey, as her maiden voyage, under charter to James Baines & Co., called for Liverpool.

Amid the huzzas of at least fifty thousand persons, who had assembled to see the interesting ceremony, with the East Boston Band playing the inspiring air "Hail Columbia" and as a cannon boomed forth its louder notes—the *Great Republic* glided from the ways as smoothly and easily as any ship that ever was launched.

Though there were some who shook their heads—old seafaring men, legitimate sons of Neptune, who firmly believed in traditions of the sea,—Captain Alden Gifford named the *Great Republic* with Cochituate water instead of the traditional bottle of Champagne. This was said to be in consequence of her builder's desire to humor the numerous Boston ladies who were then advocating temperance. However, the writer was told by Mr. E. L. Hersey, who served as Superintendent of the McKay yard many years, that his own father, Cornelius W. McKay (Donald McKay's oldest son), with some boon companions, purloined the Champagne from the mould loft the night previous to the launching. This remained undiscovered until all was in readiness for the mammoth clipper to slide off, and Cochituate (then being introduced as Boston's drinking water) had to be hastily substituted.

A pamphlet issued by Donald McKay in 1853, written

by a "Sailor" (this last cognomen being assumed by
our good friend, Duncan McLean), gave this elaborate
description of the *Great Republic:*

She is 325 feet long, has 53 feet extreme breadth of beam,
and 39 feet depth of hold, including 4 complete decks. The
height between her spar and upper decks is 7 feet, and between
the others 8 feet; and all her accommodations are in the upper
between decks. The crew's quarters are forward; and aft she
has sail rooms, store rooms, accommodations for boys and
petty officers, and abaft these, two cabins and a vestibule.
The after cabin is beautifully wainscotted with mahogany,
has recess sofas on each side, ottomans, marble covered tables,
mirrors and elliptical panels ornamented with pictures. She
has also a fine library for the use of her crew, and spacious
accommodations for passengers.

On the spar deck there are five houses for various purposes,
but such is her vast size, that they appear to occupy but little
space. She has an eagle's head forward for a head, and on the
stern, which is semi-elliptical in form, is a large eagle, with the
American shield in his talons. She is yellow metalled up to
25 feet draught, and above is painted black. Instead of bulwarks, the outline of her spar deck is protected by a rail on
turned stanchions, which, with the houses, are painted white.
Of her materials and fastenings we cannot speak too highly.
She is built of oak, is diagonally cross-braced with iron, double
ceiled, has 4 depths of midship keelsons, each depth 15 inches
square, three depths of sister keelsons, and 4 bilge keelsons,
two of them riders, and all her frames are coaged, also the
keelsons and waterways, and she is square fastened throughout.
She has three tiers of stanchions, which extend from the hold
to the third deck, and are kneed in the most substantial style.
She also has many long pointers and 10 beamed hooks forward
and aft. In a word, she is the strongest ship ever built.

The *Great Republic* had four masts, the after one, fore-and-aft rigged, named the Spanker, was sometimes called the "McKay mast" by old salts; the others had Forbes' double topsail yards.

And here it is well to state that Capt. Robert B. Forbes was the first American who applied double topsail yards to large shipping—the greatest blessing ever conferred on seamen. By adopting this rig, McKay displayed the same practical common sense which characterized all his mechanical operations.

The lower masts, commencing with the fore, were 130, 131, and 122 feet long, and lower yards 110, 120 and 90 feet long, and the other spars in like proportions. She carried nothing higher than royals forward or aft, and was very snug and strongly rigged.

She was loaded and unloaded by means of a 15 horse power steam engine stationed upon the deck (which it was claimed could be removed into a huge longboat, constructed to be used as a propeller in the tropics, by which the *Great Republic* was to be towed in calms. This engine could also be utilized for hoisting sails, pumping ship, etc.). Here was a new step in the industry of the world. On land, we then had a steam engine that would dig and load; it was called, in compliment to the most effective worker with a spade, a "Steam Paddy". This valuable instrument of labor, which worked with great efficiency, and would never "strike" for an increase of wages, was nicknamed the "Steam Tar."

The following additional facts in relation to this noble clipper are interesting:

```
Hard Pine used in construction................1,500,000 feet
White Oak.........................................2,056 tons
Iron............................................336½ tons
Copper, exclusive of her sheathing..................56 tons
Number of days' work upon her hull................50,000
Yards of canvas in a suit of sails....................15,653
Stowage capacity..............................6,000 tons
Crew.............................100 Men and 30 Boys
```

There was little timber in the *Great Republic* longer than 50 and 60 feet, but the oak was in much shorter lengths. Timber was becoming somewhat scarcer along the Atlantic seacoast, and it had grown very scant in New England, where Donald McKay and other Boston shipbuilders were compelled to secure their supply. Southern timber was finding its way plentifully to the Northern markets and at this time Southern pitch pine was introduced.

Captain Lauchlan McKay, formerly commander of the *Sovereign of the Seas*, was placed in command of this, the finest and largest sailing vessel afloat. After being fitted out, she was to proceed to New York and load for Liverpool. While this was being done, Capt. R. B. Forbes as President of the recently incorporated "Sailors'

Snug Harbor" carried on the following correspondence with Mr. McKay:—

Boston, October 6, 1853.

DONALD McKAY, ESQ.,

My dear Sir:—

As your ship, the *Great Republic*, is likely to be visited by thousands of admirers, I suggest that you make her the medium of doing a great service to an institution which is about going into operation, and of which I am, for want of a better, the presiding officer. The "Sailors' Snug Harbor of Boston" has the sympathy of all those who take an interest in ships, and they would willingly pay a "York shilling" to see your ship and at the same time serve a benevolent object. If you approve of the suggestion, I will carry it out at once by sending a competent agent on board, and if any one should by mistake drop a dollar into the purse, I will give him credit for it.

I am very truly
Your friend and servant
(signed)   R. B. FORBES.

EAST BOSTON, Oct. 8, 1853.

CAPT. R. B. FORBES:

Dear Sir:

Yours requesting my concurrence in your very benevolent suggestion, that of having the privilege of collecting a small sum from the visitors to the *Great Republic* for the benefit of the "Sailors' Snug Harbor" in Boston, has been received. I assure you that nothing will give me more pleasure than to afford you such an opportunity. This class of men have too long been neglected: they do the labor, they sail the clippers of which we boast as a nation; and any little reward that they may be able to collect in this way, will be highly pleasing to me. And I hope the public will contribute in this way, and

235

feel it to be a privilege to be able to build up a bulwark to shelter the weather-beaten sailor, now no longer able to earn his bread by his perilous profession.

I am, dear sir, yours truly

(Signed) DONALD McKAY.

This Seamen's Society had not gone into actual operation for want of funds, and the $1000 thus collected enabled them to start.

On her first passage from Boston to New York the *Republic* was accompanied by the towboat *R. B. Forbes* and was tried under her topsails and courses, and sailed well and steered easily. It has been reported that on this trip when rounding Cape Cod she sailed so fast that the *Forbes* was dragged astern of her for many miles.

The night of December 26th–27th, 1853, should long be remembered in the annals of New York City as the period when fire did its work upon the noblest specimen of naval architectural skill of which our country could boast. Upon arrival in New York, the *Great Republic* took cargo for Liverpool, and she was almost ready for sea when the disaster took place that shattered the hopes of her builder-owner. The rigging of the big clipper, which was lying at the foot of Dover Street, East River, caught fire from blazing cinders blown from some burning buildings nearby. Owing to the height of her masts it was impossible for the fire engines to play upon the flames and the consequence was that the falling spars soon set

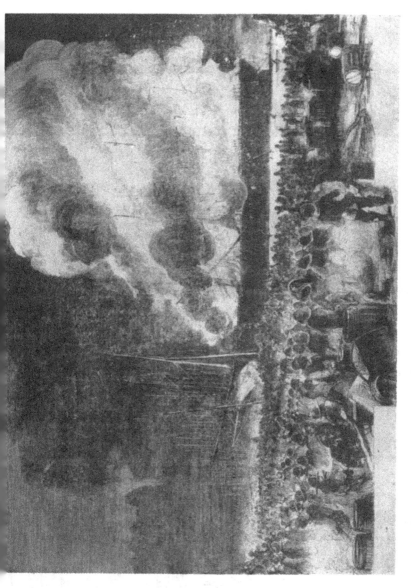

BURNING OF THE "GREAT REPUBLIC" AT NEW YORK, DECEMBER 27TH, 1853

From an old print in the McKay Collection.

DONALD McKAY'S UNEXCELLED ATTAINMENTS IN THE PRODUCTION (CALIFORNIA CLIPPERS—THE BEST AND THE MOST BEAUTIFUL SHIPS IN THE WORLD—1850 TO 1853

"STAG HOUND," 1534 Tons.
Launched, December 7th, 1850.
Estimated Cost $45,000.

"FLYING CLOUD," 1782 Tons.
Launched, April 15th, 1851.
Reported Cost $50,000.

"SOVEREIGN OF THE SEAS,"
2421 Tons.
Launched, July, 1852.
Estimated Cost $95,000.

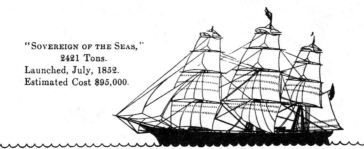

"GREAT REPUBLIC," 4555 Tons.
(The Largest Ship in the World.)
Launched, October 4th, 1853.
Building and Outfitting Cost
Approximately $300,000.

her deck ablaze. Some of the sails were bent and when they once became ignited no human power could save her. It was melancholy to see this beautiful and almost human ship, the work of months, destroyed in a few brief hours! But a few days before, she had been regarded by admiring thousands, including the Governor, Members of the Legislature, and other prominent citizens. In the same fire the clipper ships *Joseph Walker* and *White Squall* were destroyed.

The destruction of the *Great Republic* was justly considered a national calamity. Her extraordinary magnitude and the anticipations connected with her, had caused her to be regarded with something of national pride and interest. Besides, she was a scientific experiment. We were to learn from her whether the speed of ships increases indefinitely in proportion to their size, or whether builders had already reached the maximum of velocity as well as the bounds of safety in nautical construction, and no less an authority than Lieut. M. F. Maury had said that she was the fastest sailing ship afloat.

A few days before the fire Mr. McKay had refused $280,000.00 for his ship, as she then lay, ready for sea. It is estimated that he had ventured something like $300,000.00 in building and outfitting her.

Some five weeks after the fire the underwriters took over what was left of the *Great Republic*. Her builder returned from New York, having collected $235,000.00:

the sum insured upon her and her freight. Many of the underwriters and other friends gave him a public dinner, at which every encouragement was held out for him to restore the great ship to her original beauty and completeness, but he did not buy her in.

When we consider her vast size, the beauty of her model, her amazing strength and the completeness of her outfits, no sailing vessel ever designed or constructed of wood compared with Donald McKay's *Great Republic*.

THE "GREAT REPUBLIC" AS REBUILT
FEBRUARY, 1855

SALVAGED, razeed to 3357 tons, with approximately 2000 tons less stowage capacity, and under greatly reduced sailing rig, Donald McKay's masterly creation was no more! What wonders of speed might not this ship o' ships have performed as he built and rigged her!

Repeatedly it has been stated by American and English writers, that the *Great Republic's* builder never recovered from the disappointment occasioned by the loss of his ship. Long-cherished dreams were undoubtedly rent asunder and his truly ambitious plans thwarted, for he built his mammoth clipper to conquer the wind and waves, as well as for financial gain and fame. Fair-minded chroniclers, however, must admit that it was after the *Great Republic* disaster that Donald McKay won so much world-wide renown as designer and builder of the famous Australian Black Ball Fleet of Clippers—the finest and swiftest sailing vessels that ever fretted the uttermost seas with a spurning keel!

Here are some incontrovertible facts that may prove interesting even at this late date:—

On the night of the *Great Republic* fire, Mr. McKay received a telegraphic despatch informing him of her destruction. He walked the floor until daylight, the telegram in his hand, saying nothing and could not be consoled by his wife or any member of his family. Next day he went to New York, viewed the wreck and fully realized that all possible had been done to save his ship. Directly upon his return to Boston, he was to be found in his shipyard, making ready to launch the clipper *Lightning*. He was determined not to be idle, despite the calamity which had befallen him.

As before stated, Donald McKay received the sum of $235,000 from the underwriters, every policy being paid in full. All contentions that he suffered a financial collapse through the loss of his ship can therefore be set at naught.

Rebuilding the *Great Republic* occupied more than a year. To further emphasize what wonderful recuperative powers her designer and builder possessed, we will now review his shipbuilding operations during her reconstruction, say January 1854 to February 1855, when the rebuilt craft was ready to go to sea:

Ship *Lightning*	2,083 tons
Ship *James Baines*	2,526 tons
Ship *Champion of the Seas*....	2,447 tons
Ship *Blanche Moore*	1,787 tons
Ship *Commodore Perry*	1,964 tons

Ship *Santa Claus*	1,256 tons
Schooner *Benin*	692 tons
Ship *Japan*	1,964 tons
Ship *Donald McKay*	2,594 tons
TOTAL TONNAGE	17,313 tons

Within this period of thirteen months he launched eight square-riggers and one fore-and-after, aggregating 17,313 tons, as against 11,910 tons the previous year (wherein is included *Great Republic*). Only 2597 tons was turned out of the McKay yard during the next eleven months, February to December, 1855. Freights were beginning to slacken and the tide of economy was setting in, besides steam was commencing to threaten the trade of the sailing vessels.

From the foregoing it can be readily seen that for about a year *immediately following his surrender of the "Great Republic" to the insurance companies, Donald McKay produced greater tonnage and more fine ships than ever before in his career.*

The bottom had dropped out of California freights and American shipping conditions were bad, so Mr. McKay was singularly fortunate in being commissioned by an English house, James Baines & Co. of Liverpool to build the clippers *Lightning, James Baines, Champion of the Seas* and *Donald McKay.*

Furthermore, he sold the same firm the ships *Commodore Perry* and *Japan*, which he had built on speculation

241

with the funds paid him on account of the *Great Republic* insurance. It is a noteworthy fact that no shipbuilder in America or elsewhere ever built so many ships for his own account.

Now let us revert to the *Great Republic*. After being surrendered to the insurance companies, she had been bought by Captain N. B. Palmer for Messrs. A. A. Low and Brothers. He purchased her as she lay at the bottom of the East River—"as is and where is" to quote an old marine insurance term. She was rebuilt by Sneeden & Whitlock at Greenpoint, Long Island, under his personal superintendence.

She was still the largest merchant ship afloat. True, her sail plan had been cut down and changed from Forbes' to Howes' double topsail rig, and all her spars reduced about 25 per cent.—the fore- and mainmasts 17 feet, the fore- and mainyards 20 feet, and all other spars in proportion. She still carried four masts and under reduced rig required only about 50 able seamen, a dozen ordinaries and some apprentices to handle her.[1]

---

[1] The author has often been asked about the spanker or fourth mast of the *Great Republic*, some people asserting that when the ship was rebuilt it was removed and she never went to sea with four masts, often substantiating their assertion with a lithograph showing only three masts and by the printed account in Hall's Governmental Report on Shipbuilding to same effect. Francis B. C. Bradlee, in an interesting booklet "The Ship *Great Republic* and Donald McKay Her Builder," reproduces an old photograph from his collection, showing this ship at San Francisco in 1860, and in it the spanker mast is clearly shown. Mr. Bradlee also writes that "an old sea captain named Luther published a pamphlet of reminiscences and in it he mentions being a member of the *Great Republic's*

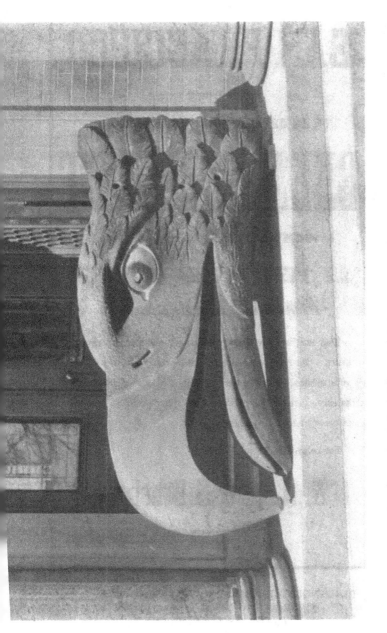

FIGUREHEAD OF "GREAT REPUBLIC"

This is one of the few specimens of Donald McKay's craftsmanship left to posterity. It reposes in the Public Library, Stonington, Conn., and is owned by Mrs. Richard Fanning Lopez, niece of Capt. Nathaniel B. Palmer. Photographed through the kindness of Mr. Thomas Whitridge Cutler, of Stonington.

FOR
SAN FRANCISCO

THE CELEBRATED CLIPPER SHIP

GREAT REPUBLIC

LIMEBURNER, Commander,

AT PIER 36 EAST RIVER,

Will have immediate dispatch.

This ship has been newly coppered, and put in complete order. Her short passages, and the perfect delivery of cargoes, entitle her to a preference with shippers. Having large hatches, she can take bulky freight under deck. Two-thirds of her capacity is already engaged.

For balance of Freight, apply to

A. A. Low & Brothers,

31 Burling Slip.

SHIPPING CARD ADVERTISING "GREAT REPUBLIC"
WHEN ENGAGED IN THE CALIFORNIA
TRADE, 1858 TO 1860

Reproduced through kindness of Miss M. Limeburner, granddaughter of Captain Joseph Limeburner of the *Great Republic*.

A carved billet head and scroll replaced her emblematic figurehead—a beautifully carved and gilded eagle's head. The original figurehead is now owned by Captain N. B. Palmer's niece, Mrs. Richard Fanning Loper of Stonington, Conn., and reposes in that town's public library. We believe it to be one of the few specimen-parts of McKay's famous American clipper fleet left to posterity.

In February, 1855, under command of Captain Joseph Limeburner, the *Great Republic* sailed on her first voyage for Liverpool, and made the run from Sandy Hook in *13 days*. Upon arrival in the Mersey, she could not dock owing to her great draft. Some three days later, at London, she was obliged to lie in the Thames, as no dock was large enough to take her.

At both ports, the *Republic*, which embodied the perfection of her type, created a sensation. Captain Limeburner often narrated, in after years, how people who came aboard would ask "whether he had left any lumber for shipbuilding in the United States or brought it all with him?"

Shortly afterwards this mammoth American clipper was chartered by the French government, then wanting transports for the Crimean War. Upon one voyage she took 1600 British soldiers from Liverpool to Marseilles,

crew (before the mast) in 1862 and that she then had but three masts. So that the spanker, or jigger mast must have been removed between 1860 and 1862." This confirms personal statements made to the writer many years ago by two former members of Captain Limeburner's crew.

where it was necessary to employ four steamships to take them up the Black Sea.

The harbor of Kamiesch (Crimean War) was densely blocked up with the transport fleet, and an enthusiastic American traveller writes thus:—

The French Flag floated from her mizen mast, but the Stars and Stripes were at the peak. Prouder than the one hundred and twenty gun ship-of-the-line of the French, the *Napoleon III*, more dignified even than the *Agamemnon* of the English, the *Great Republic* of the Americans looked in her unassuming greatness—the commander of the fleet.

The *Great Republic*, Captain Limeburner; the *Queen of Clippers*, Captain Zerega, and the *Monarch of the Sea*, Captain Gardner, were anchored side by side, and I never felt prouder of my country than in witnessing these magnificent clippers from New York, so superior to any of the transports of other nations. America never sent better representatives abroad—the peaceful messenger of commerce is always welcomed, while we only hail the ship of war as a State necessity.

Later he goes on to say:—

All the American ships are in the employ of the French Government, the English having chartered no American transports. During the dull times, the transport charters have proved a splendid business for such of our ships as were so fortunate as to get employment. The *Great Republic* must have paid for herself by this time; but the game is now up, and a thousand sail of transports will shortly find their way back to assist in deadening freights and consequently depreciating shipping property.

Though deprived of her unparalleled sailing facilities as originally designed, the *Great Republic* often, with a leading breeze, distanced the swiftest steamers, in the Mediterranean, when running between Marseilles and the Crimea.

Returning to New York, late in 1856, at the close of the Crimean War, still under Captain Limeburner's command, she was successfully employed in the California trade. On December 7, 1856, she sailed upon her first voyage to San Francisco, with a crew consisting of as "tough an aggregation of talent" as ever shipped aboard anything short of a buccaneering expedition. Captain Limeburner and his able officers always went armed; it is said that the top-gallant sails were never clewed up during the passage, and Cape Horn was rounded with skysails set, in 45 days and some hours. Her motley crew proved good sailors, for the *Republic* made the quickest California run of the year, 92 days. The fifth day out from Sandy Hook she logged 413 miles; crossed the line in 15 days, 18 hours, the fastest sailing time on record. If she had not been delayed within 500 miles of the Heads for five days in calms and fogs, we believe the *Flying Cloud's* record would have been eclipsed. It was on this record passage that she beat the *Westward Ho!*, and the New York men who had backed her in the race won large sums from the *Westward Ho!'s* many Boston admirers.

From San Francisco to Callao went the *Great Republic;* then loaded guano at the Chincha Islands and sailed for London. She shipped a tremendous sea shortly after rounding the Horn which stove her deck and did other damage, so she put in at the Falklands. She was laid up at these Islands about six months. Captain Limeburner was obliged to send to the mainland for provisions, once as far as Montevideo for some supplies and lumber to make repairs. Not until January 1858 did she finally arrive at London.

During the next two years the *Republic* travelled principally between New York and San Francisco.

A reaction in American shipping circles had set in. The California excitement was over, the rush to the gold mines virtually slackened, and the settlers of the new territory were producing for themselves those common necessaries of life which had previously been sent to them from the East by clipper ships. Owing to enormously high prices of food, agriculture had made rapid progress and California now raised great quantities of grain. Early in 1861, with one of California's first large grain exportations the *Great Republic* left San Francisco for Liverpool, arriving there in 96 days.

At New York, from Liverpool, shortly after the Civil War broke out, she was seized as rebel property, it having been discovered that the majority of her owners were Southerners. However, this was soon adjusted by A. A.

Low & Bros. who still retained an interest in her. In 1862 she was chartered to the United States Government for the transport of troops, and following a passage to Port Royal and return, she assisted in the transportation of General B. F. Butler's troops to Ship Island.

These troops mutinied off Port Royal, but eventually the outbreak was suppressed and they reached their destination. She was then sent down with coal to the Southern Squadron, and while bunkering two United States gunboats, she broke adrift during a gale and went ashore in the Mississippi. She was salved with little difficulty—proving the advantageousness of a vast surface of floor with a dead rise of only about 20 inches. She continued in the Governmental service a few months, then returned to New York.

The *Great Republic* now resumed her part in the California trade. Humdrum voyages to and from Pacific ports followed, forming part of this—her after-career. Captain Josiah Paul succeeded "Joe" Limeburner, who had commanded this vessel from the recommencement of her career.

A. A. Low & Brothers still owned her and in 1867 she was sold, after being laid up for about two years in New York, where she attracted a lot of attention, especially from foreign visitors.

In January 1869, she was sold to the Merchants Trading Company of Liverpool and renamed *Denmark*. After her change of ownership (and flag) she was generally employed in the East India Trade, carrying various cargoes until 1872.

Her final end was a watery grave off Bermuda. The ship in a strained and run down condition was caught in a hurricane and sprang a leak, while on her way light from Rio de Janeiro to St. John, N. B., where she was to be repaired and then take a cargo of lumber to Liverpool. All hands reached land in safety.

For some time afterwards it was thought she was still afloat and might be picked up. The *Great Republic's* career certainly proved one of more than ordinary eventfulness.

CHAPTER XXVII

FIGURATIVE language could hardly conceive a more ingenuous nomenclature for an American Clipper Ship than the *Romance of the Seas*.

This ship registered 1782 tons, when measured by the Custom House officials. She was 240 feet long on deck, had 39 ft. 6 in. breadth of beam, depth 29 ft. 6 in., including 8 feet height of between decks. Her lines were concave and her ends very long and very sharp. Indeed she was the sharpest clipper of her size ever built at the McKay Yard and was expected to beat the celebrated clipper ship *Flying Cloud*, which still headed the list of California passages.

In model she was a beautiful vessel. Although concave below, she was convex in the lines of her upper works, and her outline on deck was true as the sweep of a circle. She had 15 inches dead rise at half floor, about 6 inches rounding of sides, and between 4 and 5 feet sheer, which was graduated her whole length, and nearly alike at both ends. For a figurehead she had a small female

249

figure, intended to represent Romance, with the name of Scott on one side, and Cooper on the other—the greatest romancers of the century. The stern was nearly semi-elliptical in form, and rose from the line of the planksheer, the molding of which formed its base. It was very light and graceful and tastefully ornamented with gilded carved work.

The *Romance of the Seas* was the last extreme clipper ship built by Donald McKay for the California trade. With extremely fine lines, she was a beautiful vessel and like every craft this master mechanic built for her owner, George B. Upton of Boston, no vessel was more thoroughly constructed, or more smoothly finished. Although heavily sparred, such was the care that had been bestowed in balancing her spars, she proved an exceedingly fast ship, especially in moderate weather.

The application of double topsails to large shipping was the greatest improvement of all time in the rig of square-rigged vessels, and to Captains Forbes and Howes (rival claimants for its introduction), the men of the sea owe a debt of eternal gratitude. Only those who followed the sea when single topsails were common can appreciate the inestimable value of double topsails. Previous to their use, seamen were constantly exposed to instant death in reefing the sails. Double topsails can be carried to the last minute in squally weather, because the upper topsails can be lowered and lie becalmed before the lower

DONALD McKAY IN THE ZENITH OF HIS FAME

Age about 45 years.   From a lithograph in the author's possession.

CAPTAIN JOSIAH PERKINS CREESY.
Of the *Flying Cloud*.

CAPTAIN LAUCHLAN McKAY.
Who commanded *Sovereign of the Seas* and was to take the *Great Republic* out on her maiden voyage.

MATTHEW FONTAINE MAURY.
"The Pathfinder of the Sea."

CAPTAIN PHILIP DUMARESQ.
Commander of *Bald Eagle* and *Romance of the Sea*.

JOSEPH LIMEBURNER.
Who long commanded the *Great Republic*.

SOME OF THE SHIPMASTERS WHO MADE DONALD McKAY'S CLIPPERS FAMOUS AN THE GREAT MARINE SCIENTIST WHO MADE THEIR ACHIEVEMENTS POSSIBLE

ones until a squall has blown over, and be reset without requiring a man to go aloft. Donald McKay's *Great Republic* originally carried Forbes' rig; the *Romance* and every other ship he afterward constructed carried topsail yards and all the improved rigging appliances invented for use on sailing craft.

The *Romance of the Seas* loaded with despatch in Messrs. Timothy Davis & Co.'s line of San Francisco Clippers, then located on the South side of Long Wharf, Boston. Captain Philip Dumaresq commanded her, and it is not too much to say that as an accomplished, daring and successful shipmaster, he had few equals.

Sailing upon her first voyage, December 16, 1853, she arrived at San Francisco, March 23, 1854, having made the passage in 96 days. The crack New York clipper *David Brown*, facetiously described by Boston newspapers as "built to beat the world and the rest of mankind," sailed from her home port three days and some hours in advance of the *Romance* from Boston, and the latter ship arrived out, anchored and had her sails furled before the *David Brown* came to anchor.

Off the coast of Brazil the two vessels were close to each other, and during the remainder of the voyage were never over 150 miles apart, sometimes only 40, each experiencing the same weather, each losing a jib-boom in the same gale, and each arriving at the common destination at almost the same hour.

There with great despatch cargoes were discharged, ballast was taken in, and new crews were shipped, so in eight days both ships left the harbor of San Francisco side by side for China. The *Romance* forty-five days afterwards anchored in Hong Kong about an hour first— never having taken in her main skysail from the time of her leaving San Francisco.

In this age of steam and enterprise, it is difficult to realize the intense interest with which these clipper ship Races were regarded. Many bets were made, especially by Boston with New York shipping men, the *Romance* being the favorite. No branch of mercantile activity today inspires so much wholesome sporting interest and intelligent enthusiasm as did those ocean-to-ocean sailing matches by American clippers that outsailed everything on the seas from 1850 onward for many years.

The *Romance* loaded at Whampoa with tea and was 102 days to London, while the *Brown* sailed from Shanghai to London and took 111 days. At this time clipper ships and their handling had reached a high standard of excellence, so it is to be regretted they had no further opportunity of sailing together.

From London the *Romance* went out to Hong Kong, Captain Dumaresq having relinquished command, and then returned to England from China after a very slow passage. Afterwards she sailed to Boston, where she was outfitted for a trip around the Horn, sailing from New

York July 3, 1856, under command of Captain Henry. Arrived at San Francisco October 24th, 113 days thereafter, having encountered heavy gales off the Cape, which carried away her impressive figurehead and did other damage.

A record passage of 34 days, 4 hours, San Francisco to Shanghai, followed, after which the *Romance* was engaged in the China trade for some little time.

With California grain cargoes to England, she made one or two quick runs, and then embarked again for the Far East. At Hong Kong on December 31, 1862, she loaded for San Francisco and was never heard of afterwards.

As long as ships last, and as long as men go down to the sea in ships, our California and China clippers will never be forgotten. Built at a time when speed was of more consequence than cargo-carrying capacity or comfort, these vessels raced home to New York or Boston and to England with tea and other freight from the Orient, and were driven hard to take well-paying cargoes and new settlers out to the California gold fields. The *Romance of the Seas* was the last of the fleet of real out-and-out Yankee clippers that Donald McKay built in the halcyon days of American shipping.

The prosaic iron steamship that has succeeded the clipper on the trade routes, although more efficient, will never be eulogized so fascinatingly as this:

O, fair she was to look on, as some spirit of the sea,
When she raced from China, homeward, with her freight of
    fragrant tea;
And the shining, swift bonito or the wide-winged albatross
Claimed kinship with the clipper beneath the Southern Cross.

From the haven of the present she has cleared and slipped
    away,
Loaded dead and running free for the port of yesterday,
And the cargo that she carried, ah! it was not China tea,
She took with her all the glamor and romance of life at sea.

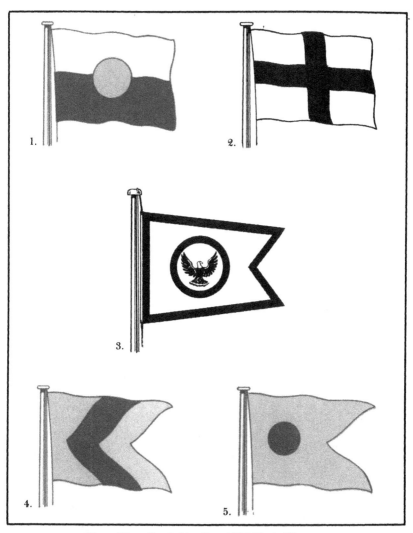

House Flags Carried by Donald McKay's Clippers.

1. Messrs. Sampson & Tappan, Boston. *Red circle with white stripe above and blue stripe below.*
2. George B. Upton, Boston. *Blue cross on white field.*
3. Donald McKay, Boston. *Blue eagle, circle and border on white field.*
4. Grinnell, Minturn & Co., of N.Y., California Line. From left to right: *Red stripe, blue stripe, yellow stripe.*
5. James Baines & Co., Liverpool, England (Australian Black Ball Line). *Black circle on red field.*

*In the original edition, where the plates were in color and bore no verbal descriptions, the red used was of a very orange quality. Unsure of what was intended, we say "red" here for brevity.

PART IV

AUSTRALIAN CLIPPERS, 1854–1856

CHAPTER XXVIII

W HAT might be termed "America's Golden Age of Sail" was now ended. Clipper ships, of the extreme type, with every element in them made subservient to speed, could only exist with high rates of freight, and freights were brought down on the run. The rush to California was over.

While it lasted it certainly was the most interesting period in the history of American shipping, and the glory and romance of our "Cape Horn Clipper Fleet" will never fade as long as a love of the sea exists.

Donald McKay's fame attracted the notice of James Baines & Co., of Liverpool, who contracted with him to build several first class ships for their Australian Black Ball Line. It is the only time in history that Great Britain bought not one but a fleet of ships outside of her own domain.

The wonderful performances of this fleet of splendid record-breakers—the *Lightning* and *James Baines* were the most famous—was certainly a befitting aftermath to that glorious period when our California clippers broke record after record racing around the Horn. So well satisfied were Messrs. Baines & Co. that they wrote to Mr. McKay that he had given them in their ships "all and more than they had asked or expected."

Shuttling back and forth between England and the island continent, these Boston-built ships, sailing under the British flag, showing superior qualities as seaboats and carriers, greatly contributed towards the world-wide fame of American shipbuilding. Curiously enough America then suffered that decadence which eventually drove her shipping off the seas.

Lightning, The Fastest Ship that Ever Sailed the Seas. Her record of 436 miles in 24 hours remains unequaled. From painting by Anton Otto Fischer. Courtesy of George R. Gorman, Esq.

CHAPTER XXIX

"LIGHTNING," 2083 TONS
LAUNCHED JANUARY 3RD, 1854

"No timid hand or hesitating brain gave form and dimensions to the *Lightning*. Very great stability; acute extremities; full, short midship body; comparatively small deadrise, and the longest end forward, are points in the excellence of this ship."—John Willis Griffiths in the *Monthly Nautical Magazine*, August, 1855.

AND now gold has been discovered in Australia, so there's another rush by sea for that Land of Promise! The contest between the United States and Britain had assumed a new aspect, for English ship-owners, to retain the increasingly valuable Australian trade, must perforce buy American or Canadian-built ships.

One of the first shipping men in England, who foresaw the advantage of owning fast American-built clipper ships, was James Baines of Liverpool, whose Black Ball Line, sailing between that port and Melbourne, Australia, had most successfully contended against the keen rivalry in the Australian trade, especially from London shipping and mercantile interests which had for years practically enjoyed a monopoly of that business until Baines and other enterprising Liverpool shipping men revolutionized existing conditions.

The house of James Baines & Co. determined to own the finest and fastest ships that could be built, and decided, despite much criticism and national ill-feeling in England, to have them constructed in America by Donald McKay. Mr. Baines, with a wonderful eye for ships, had long admired from afar the beautiful specimens of McKay's handicraft that came under his observation; and now, with success achieved, he was ordering from the Boston clipper designer-builder three of the finest and largest ships in the world! Another large vessel, which he generously named the *Donald McKay*, also found its way on the stocks for his account. Furthermore two vessels in process of construction at McKay's yard, the *Japan* and *Commodore Perry*, were added to the Australian Black Ball Fleet.

Shipping men on both sides of the Atlantic looked on aghast at the immensity of Baines' transactions with America's premier shipbuilder. Undoubtedly in the coming together of James Baines and Donald McKay there was consummated a union of genius by which the best of America and of Britain wrought together for the fame of ships of sail before steamers had assumed ascendency upon the ocean.

"Jimmy" Baines, as he was familiarly called in Liverpool shipping circles, had achieved meteoric success. He

was a lively little man, fair with reddish hair, and sparkling eyes, whose manner told of youthful enthusiasm and
maturer wisdom in unaccustomed blending. Of an eager,
generous disposition, he was far from being the cool,
calculating business man usually allied to commercial
success. He branched out in 1851 as an important factor
in the Australian emigrant trade when he purchased a
big St. John-built timber ship, fitted her in a rather
elaborate manner for emigrants and placed her under
command of James Nicol Forbes—famous in nautical
circles as "Bully" Forbes,—who certainly made the "old
hooker" travel! This was the *Marco Polo*—the first ship
to bring fame to Baines' Black Ball Line, and Baines did
so well with her in the Australian trade that before very
long he had a number of ships running on it.

Baines had in 1853 operated McKay's *Sovereign of the
Seas* successfully. To attract shippers, he dramatically
offered space in her hold at seven pounds per ton, the
shipper to recover fifty shillings if she did not beat every
steamer on the run to Australia. And Baines kept the
money!

"Bully" Forbes was given command of the *Lightning*
and he came from England to superintend her outfits at
McKay's shipyard.

She was the first of that historic American clipper
quartette that, notwithstanding transfer from the Stars
and Stripes to the Union Jack, shed luster upon both, and

rendered wonderful service to England and Australia's commercial interests and their peoples' welfare. McKay's *Lightning* has the distinction of being the first vessel ever designed and built in this country for an English house,— and she was the largest ship belonging to Liverpool upon her arrival there.

This splendid vessel was 226 feet long at the load-displacement line, and 243 over-all, from her knight-heads to the taffrail; had 44 feet extreme breadth of beam, 23 feet depth of hold, including $7\frac{1}{2}$ feet height between decks; her rise of floor was 20 inches and her sheer $4\frac{1}{2}$ feet, which was graduated her whole length, and rose gracefully towards the ends. She had sharper ends than any clipper ever built in the world, and her lines were decidedly concave. At the load-displacement line, a chord from the extreme of the cutwater to the rounding of her side, would show a concavity of 16 inches, the curved line representing the segment of an ellipsis. The stem raked boldly forward, and the bow flared as it rose, but preserved its angular form to the rail, and was there convex in its outline. Completing the fair outline of the bow, was a beautiful full length figure of a young woman, with flowing drapery and streaming hair, holding a golden thunderbolt in her outstretched hand. This was her only ornament forward, for she had neither head nor tail boards, nor any other appendage for the sea to wash away.

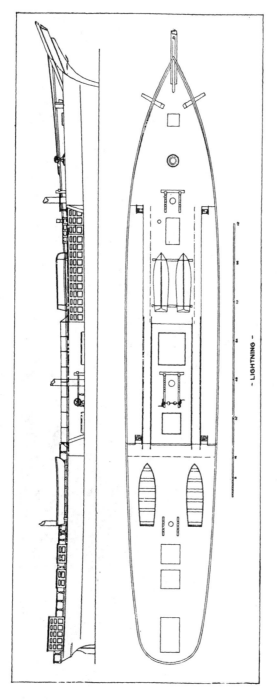

DECK PLAN OF PACKET "LIGHTNING"

Drawing made by Alfred S. Brownell, from description of this famous ship in *Boston Atlas*, January 31st, 1854.

IN THE ROARING FORTIES

The *Lightning* and *James Baines*, the fastest sailing vessels the world has ever known, making records between England and Australia. Reproduced through the courtesy of the Battery Park Branch of the Bank of America, New York, from painting by Charles Dixon.

Aloft the *Lightning* was heavily and strongly rigged. Her main yard was 95 feet in length, and the total height from her deck to the main skysail truck was 164 feet; her lower studdingsail booms were 65 feet in length; her topsails and topgallantsails were diagonally roped from clews to earings, and her fore and main stays, lower rigging, and topmast stays and backstays were of 11½ inch Russian hemp, with the rest of the standing rigging in proportion. Indeed, her masts and spars were as strongly secured as skill and labor could make them.

The quarter-deck was 90 feet long, flush with the top of the bulwarks, and protected by a mahogany rail on turned stanchions of the same wood. She had also two large deck-houses, which, together with the between-decks, gave ample passenger accommodations. The quarters for the steerage passengers were comfortably fitted and well ventilated, while the saloons, staterooms, bathrooms, and smoking-room for the cabin passengers were superbly decorated and furnished.

Captain Forbes, a worthy and energetic Scotchman, regarded as a devil at sea but a pious (?) man ashore, found a congenial spirit in Captain Lauchlan McKay, formerly commander of the *Sovereign of the Seas*, who consented to accompany him upon the *Lightning's* maiden crossing. They were attended down Boston harbor by a select party of brethren and sisters of the M. E. Church, who at parting gave them their blessing. We can imagine, better than express, the feelings of "Bully" Forbes' "unregenerated crew" at this parting scene.

Here's how Duncan McLean, in the *Boston Atlas*, masterly described the beginning of this wonderful sailing craft's maiden voyage:

Not a ripple curled before her cutwater, nor did the water break at a single place along her sides. She left a wake straight as an arrow, and this was the only mark of her progress. There was a slight swell, and as she rose, one could see the arc of her forefoot rise gently over the sea as she increased her speed.

It was on Saturday afternoon, February 18, 1854, that the *Lightning* sailed from Boston bound to Liverpool. At 2 o'clock she hove her anchor up and at 3 o'clock discharged pilot off Boston Light.

This extract from the magnificent clipper's abstract log, shows what was happening aboard her on that memorable day in March, 1854.

March 1.—Wind S., strong gales; bore away for the North Channel, carried away the foretopsail and lost jib; hove the log several times, and found the ship going through the water at the rate of 18 to 18½ knots per hour; lee rail under water, and the rigging slack; saw the Irish land at 9:30 P.M. *Distance run in the twenty-four hours, 436 miles.*

A phenomenal run! The longest authenticated distance ever left astern by a sailing vessel in twenty-four hours. It entitles McKay's *Lightning* to the proud distinction of being *the swiftest ship that ever sailed the seas!*

Not until some thirty years afterward did an ocean steamship exceed the *Lightning's* fastest day's run.

"Carried away the foretopsail and lost jib,"—shows it certainly must have been blowing "great guns," for these were stout new sails, but Forbes always did carry on in the most daring manner.

From her abstract log, it appears she made the run from Boston Light to Eagle Island (coast of Ireland) in 10 days, and to the Calf of Man in 12 days, thence to Liverpool in 13 days, 20 hours—one of the shortest passages made across the Atlantic in the sailing ship's most progressive era.

At Liverpool it is claimed, £2,000 was spent on *Lightning's* furnishings and decorations below, in addition to her original cost, about £30,000.

On her first voyage from Liverpool to Melbourne, the *Lightning* did no better than McKay's *Sovereign of the Seas*—77 days; but on her return passage she hung up the record of 64 days, making a run of 3722 miles in ten consecutive days and showing 412 miles for her best day's work. The quickest trip ever made before this was in 75 days. On this passage she carried $5,000,000 in gold and dust. The voyage out and home occupied five months, eight days, including three weeks detention in Melbourne.

Upon the *Lightning's* maiden passage, Forbes originated the slogan—"Melbourne or Hell in sixty days!"

The passage from Liverpool to Melbourne at this period was considered the longest which occurred in a direct passage between any two ports of equal commercial importance, passing through 200 degrees of longitude and 124 of latitude.

Captain Forbes leaving to take charge of the ill-fated *Schomberg*, he was succeeded by Anthony Enright at the unprecedented salary of £1,000 per annum. He commanded the *Lightning* for four voyages, from January, 1855, until August, 1857, and under Enright she became one of the most popular passenger carriers in the Australian trade.

It was on his second passage to Melbourne that the "wood butchers of Liverpool," as Donald McKay styled them, were allowed to fill in the concave lines of this splendid clipper's bow with slabs of oak sheathing. A letter from Mr. McKay, which appeared in the *Scientific American* on November 26, 1859, reads as follows:

Although I designed and built the Clipper Ship *Lightning* and therefore ought to be the last to praise her, yet such has been her performance *since Englishmen learned to sail her* that I must confess I feel proud of her. You are aware that she was so sharp and concave forward that one of her stupid captains who did not comprehend the principle upon which she was built, persuaded the owners to fill in the hollows of her bows. They did so, and according to their British bluff notions, she was not only better for the addition, but would sail faster, and wrote me to that effect. Well, the next passage to Melbourne, Australia, she washed the encumbrance away on one side, and when she returned to Liverpool, the other side was also cleared away. Since then she has been running as I modelled her. As a specimen of her speed, I may say that I saw recorded in her log (of 24 hours) 436 nautical miles, a trifle over 18 knots an hour.

There was some doubt cast upon the *Lightning's* day's run and her builder certainly felt warranted in stating that he saw *436 nautical miles* recorded in her log.

But let us resume our course, running in the Australian trade. Nearly all records were made on the outward passage, England to Australia, where Baines' clippers not only carried large crews but frequently doubled them by men working their passage to the Victorian goldfields. Returning home speed was not so important, so they managed with a much smaller crew. As was so frequently the case in California, soon as the anchor was down in Australian waters, off went the crew to the "diggin's!" The ship would generally take in her cargo manned by her afterguard and perhaps the apprentices. Before her time to sail came around, there were generally disillusioned gold miners, glad of a chance to work their passage home. In the beginning of the rush, they stood out for Australian rates of pay, but it was not very long before crews were easily secured for homeward bound ships.

Three years after her wonderful maiden voyage the *Lightning* was to come within six miles of her own record. When running her easting down, bound to Australia in

March, 1857, the second greatest day's run of 430 miles was made by McKay's flyer, still serving as a Black Ball liner under command of Captain Enright.

Together with the *James Baines* and *Champion of the Seas* the *Lightning* engaged in the transportation of English troops to India during the Sepoy Rebellion. In August, 1857, she sailed with the 7th Hussars, and prior to sailing was thrown open to public inspection at Gravesend. Captain Enright having given up command, he was succeeded by Captain Byrne. The number of American-built clippers engaged in this Indian Mutiny trooping naturally caused great interest in their several passages, as also in the passages of the various British clippers and steamers.

The arrival out of the *Lightning* in 87 days sustained her previous reputation for speed. We annex a comparative table of the passages of the various vessels:

Passage of the *Lightning*   87   days
Passage of the *James Baines*   103   days
Passage of the *Champion of the Seas*   101   days
Passage of other sailing vessels.   120   days
Average passage of full power screws. . .   83   days
Average passage of auxiliaries . . . .   96¾ days

By the above table it will be seen that the *Lightning* not only beat handsomely all the other American and British-built clippers, but she had also beaten all the average passages of the auxiliary screw steamers, and came within a few days of the passages of the full powered screw steamers. This was another splendid triumph for American naval architecture.

About 1859 the *Lightning* returned to the Australian trade, in which she remained ten years. In 1867, Thomas Harrison of Liverpool purchased her; and two years later this celebrated clipper met her end by fire, being burnt at Geelong, Melbourne Harbor, October 31st, 1869, and scuttled at her anchorage, when fully loaded with wool, copper ore, tallow, leather, etc., for her homeward passage.

What the speed of this fast clipper meant is better understood when it is realized that no steamship afloat at the time could have come within a hundred miles of *Lightning's* 436 mile log for twenty-four hours,—in fact, there are few steamships outside of the great passenger liners today that can better that mark.

SOME FAMOUS SAILING SHIPS

NOTE

Copy of a newspaper clipping that appeared in *Sea Stories* magazine August, 1928. The article was sent to them by a correspondent in Australia but he did not mention the name of the paper from which it was taken.

For some days difficulty has been experienced in operating the dredge *Thomas Bent* off the Yarra Street pier, owing to the buckets striking some hard objects, and occasionally bringing to the surface pieces of timber, copper, iron and chains. To ascertain the cause of the obstruction, the Harbor Trust's diver—W. Mackinlay—made an inspection to-day, and found that the dredge was operating against an old wreck. The diver found the definite form of an old vessel—some of the timbers being copper-lined and in an excellent state of preservation—anchor chains, irons, and a mass of wreckage.

The opinion is that in dredging out the berth and approach to the pier the dredge has uncovered the remains of the famous clipper ship *Lightning*, which was burnt off the Yarra Street pier on Sunday, October 31, 1869. The ship had brought out several lots of immigrants. She loaded wool, leather and general cargo at the pier, and was almost fully loaded when she caught fire. Fears were entertained for the safety of the wharf, and the burning ship was taken off for a short distance. The artillery was called out with the intention of sinking the ship by gunfire, and she finally sank, enveloped in flames. Some of the wreckage has been brought ashore and is arousing much interest.

BURNING OF THE "LIGHTNING" AT GEELONG, MELBOURNE HARBOR, ON OCTOBER 31st, 1869

From a painting by Lars Thorsen, in possession of the author.

CHAPTER XXX

THE celebrated passage of the *Lightning* and the completeness and beauty of the ship herself, gave Donald McKay a reputation in England, second only to that which he enjoyed in America.

But it was not in the construction of clippers alone that this master-builder of ships excelled. The Winter of 1853, the most terrible on the Atlantic for many years, proved the excellence of his packets also. Of all the vessels which he built for Train & Co.'s Boston and Liverpool line, various New York packet lines and others engaged in trans-Atlantic shipping, all crossed in safety.

In beauty of model, strength of construction and some other elements of perfection, the *Champion of the Seas* was a decided improvement upon the *Lightning*. She was a three-decked ship, fitted and rigged in the most costly manner. Her ends were as long, though not so sharp or concave, and were even more beautiful in their form. She was 238 feet long on the keel and 252 feet on deck between perpendiculars, which, as the sternpost was upright, gave her a fore rake of 14 feet; her extreme breadth of beam was 45½ feet, and depth 29 feet; dead-rise at half floor, 18 inches; rounding or swell of sides, 10 inches, and sheer, 4½ feet. Her greatest breadth was precisely at the centre of the load displacement line, and she was rather fuller aft on that line than she was forward.

The full figure of a sailor, with his hat in his right hand, and his left hand extended, ornamented the bow. It was certainly a most striking figurehead, the tall square-built mariner, with dark curly hair and bronze clean-shaven face; and quoting from Clark's *Clipper Ship Era*, we continue:

A black belt with a massive brass buckle supported his white trousers, which were as tight about the hips as the skin of an eel, and had wide, bell-shaped bottoms that almost hid his black polished pumps. He wore a loose-fitting blue-and-white checked shirt, with wide rolling collar, and black neck handkerchief of ample size, tied in the most rakish of square knots with long flowing ends. But perhaps the most impressive of this mariner's togs were his dark-blue jacket, and the shiny tarpaulin hat which he waved aloft in the grip of his brawny tattooed right hand. The only exception that one could possibly take to this stalwart sailorman was that his living prototype was likely to be met with so very seldom in real life.

The *Champion* had a waist of narrow strakes, defined between the mouldings of the upper wale and the plank-sheer, and this was continued around the stern, which was semi-elliptical in form, and was ornamented with the

Australian coat-of-arms. The run was long and clean and blended in perfect harmony with the general outline of the model. Broadside on she had all the imposing majesty of a ship-of-war, combined with the airy grace of a clipper. Outside she was painted black, and inside white, relieved with blue waterways, which were the regulation Black Ball colors.

Her dining saloon, cabins, and all staterooms, everything designed for the accommodation of passengers, were the most perfect in every particular. She had a spacious topgallant forecastle fitted for the crew and staterooms for the officers.

There was about 12,500 yards of canvas in a single suit of her sails. Aloft, the harmony of her masts and yards was complete, making her a perfect picture to the eye.

After fitting out at Grand Junction Wharf, East Boston, the *Champion* was towed to New York by Boston's well-known screw propeller *R. B. Forbes* and there loaded for Liverpool.

Her first Commander was Captain Alexander Newlands who had been sent over from Liverpool by James Baines & Co. to superintend her construction. He proved so agreeable, as well as a capable coworker, designing much of the interior arrangements of this ship, that Mr. McKay formed a warm attachment for him that lasted many years.

The *Champion* went across during June–July, 1854, in 29 days, a rather disappointing Atlantic passage and during which, time after time, all her immense canvas yearned for a capful of wind.

On her first voyage to Australia, she arrived out in 75 days and home in 84. Her second and third outward runs were made in 83 and 85 days respectively. She was more remarkable for repeated good passages than for epoch-making runs, and continued in popularity as a freight and passenger carrier for about three years.

At the time of the great Sepoy mutiny the *Champion of the Seas* and *James Baines*, sister ships, and the *Lightning*, were employed by the English Government for transporting troops. The *Champion* and *Baines* were visited and examined by Queen Victoria and the Royal family before sailing from Portsmouth, England, with about 1,000 troops for India. Her Majesty expressed surprise and admiration, and was much pleased that such a fine large ship was within her domain. Both ships sailed August 8, 1857, with as many troops from the same place; the *Lightning* left some seventeen days later. Although all three engaged in a contest, the race between the *Champion* and the *Baines* proved the exactness of their similarity as regards sailing qualities, though in construction and model they differed. They both arrived at the Sand Heads within a few hours of each other, having taken 101 days to make the passage. The

Lightning made the run in 87 days; the two rival clippers encountered light winds during the early part of their passage. The season was most unfavorable for navigation of the Bay of Bengal. Some ships were obliged when nearing Calcutta to let all anchors out at night to barely keep what headway they could make in the daytime, and some of them lost as much as eleven days before they could get a pilot. Many vessels were within a hundred miles of each other during the greater part of the passage, but did not know it until they compared logs at Calcutta.

After England put down this rebellion the *Champion* resumed her sailing between Liverpool and Australia, until, after many years of service, she was taken off that line, and used only for general cargo carrying.

When homeward bound in 1876, after twenty-two years valiant sea service, the *Champion* foundered off Cape Horn.

CHAPTER XXXI

"At times I muse, at times I think,
   Of Ships of other days!
Ships of renown, ships I have seen,
   Or sailed on in untracked ways."
         (*Ships and Ways of Other Days*, by N. S.)

SPREADING abroad to the breeze some thirteen thousand yards of canvas, sparred heavily to successfully contend against even the "Roaring Forties,"—we now have to recount the trans-Atlantic feat of the McKay built clipper *James Baines*,—Boston Light to Rock Light, Liverpool, twelve days and six hours! A record run, made on her maiden passage, in September, 1854, under command of Charles McDonnell, who, like Captain Forbes of the *Lightning*, left the *Marco Polo* to take charge of this new Donald McKay creation.

There was only a slight difference between the lines of this vessel and the *Champion of the Seas*, the *Baines* having a somewhat more raking stem, which brought her

276

"JAMES BAINES"

The McKay-built Clipper that made a record run across the Atlantic, then hung up 133 days for a round-the-world passage, and is credited with logging 21 miles in one hour.

From an illustration in Clark's *Clipper Ship Era.*

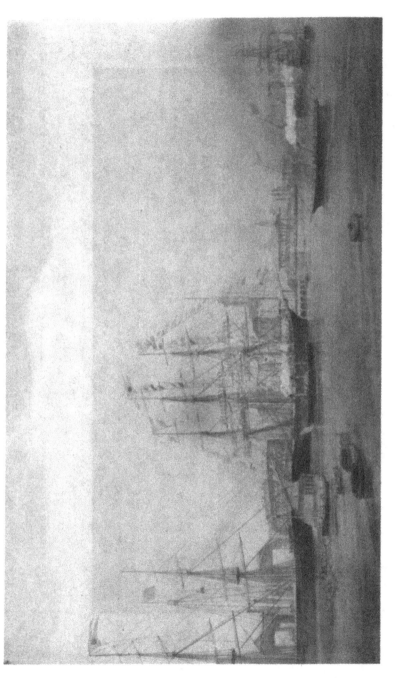

"JAMES BAINES" AND "CHAMPION OF THE SEAS"

Visit of Her Majesty Queen Victoria to these two American-built ships at Portsmouth Dockyard, before their departure for India with troops during the Sepoy rebellion. Both vessels sailed the same day and arrived together, in one of the greatest ocean races ever recorded in

lines out forward a trifle longer and sharper above the water line, so they were generally classified as "sister ships." A bust of James Baines, with whiskers and top hat complete, served as her figurehead. It was carved in Liverpool, Mr. Baines going time after time to sit for it.

Upon arrival in the Mersey, after her record trip across the Atlantic, the *James Baines* was adjudged the finest sailing ship Liverpool had ever seen.

Sailing on December 9, 1854, she made the passage from Liverpool to Melbourne in 63 days—an unbeatable sailing record. Her best twenty-four hours run was 423 miles, this with main skysail and stunsails set. Captain McDonnell, in his account of the passage to her owners, wrote "Had I only had the ordinary run of winds I would have made the voyage in 55 days."

The following extract from one of George Francis Train's *American Merchant* Letters (although erroneously giving "65 days" as passage time), portrays events that startled Australia, when as a young colony, she had to look to the sailing vessels of the "Black Ball" and "White Star" lines for her mails, etc., etc.:

MELBOURNE, Feb. 15, 1855.

The town was fairly thrown on its beam ends by the startling announcement that the *James Baines* had arrived from Liverpool with the December mails, after the astonishing and unprecedented run of sixty-five days! Can anyone now doubt Donald McKay's supremacy upon the ocean? I fancy not, for the log records of the *Flying Cloud*, the *Lightning* hence to

Liverpool in 63 days, and now the *Baines* out here in 65, will very quickly settle the question. The passages are truly wonderful, and I maintain that Donald McKay has done more to advance the science of shipbuilding than any other man. He stands the victor, and is always first in the clipper ship race ground. Clipper ships may depreciate, and over-trading in such property may prove disastrous; but the genius of the mechanic and the boldness of the man who has launched such a leviathan as the *Great Republic* will live so long as great deeds continue to be recorded.

The *Baines* came home in 69½ days, thus completing the voyage to Melbourne and back, equivalent to sailing around the globe, in the record time of 133 days. This is the best authentic sailing ship round-the-world passage.

Her second voyage out to Australia was made in 78 days, returning in 95 days; third voyage out 75 days, return 77 days. These were all her Australian voyages. In 1856 her log records:

June 16. At noon sighted a ship in the distance ahead; at 1 P.M. alongside of her; at 2 P.M. out of sight astern. The *Baines* was going 17 knots with main skysail set; the *Libertas*, the other ship, was under double-reefed topsails.

June 17. Lat. 44 S., Long. 106 E., ship going 21 knots with main skysail set.

According to the late Captain Arthur H. Clark in his "Clipper Ship Era," this appears to be the highest rate of speed ever made by a sailing vessel of which any reliable record has been preserved.

During the Sepoy mutiny the *James Baines* was among

the ships chartered by the British government to carry troops to India. On July 30, 1857, this celebrated clipper went to Portsmouth for troops, and on the 8th of August she sailed for Calcutta.

The following article, taken from a Liverpool paper, gives a very graphic account of a visit of Queen Victoria and her royal party to the *James Baines* and the *Champion of the Seas:*

ROYAL VISIT

(From the *European Times*)

The clipper-ships *James Baines*, Captain Mc'Donnell, and *Champion of the Seas*, Captain McKirdy, of the Liverpool and Australian Black-Ball Line, belonging to Messrs. James Baines & Co., arrived at Portsmouth on Monday morning from Liverpool. No ships that ever entered Portsmouth harbor created so much curiosity among men-of-war's men as these great merchantmen. High and low have been on board to visit them, and the Port Admiral, Sir George Seymour, expressed his unqualified astonishment at examining the speed logged by these mercantile clippers. The *James Baines* will take in nearly 1,000 of the 97th and other troops, and the *Champion of the Seas*, a like number of the 20th foot and other regiments, on Thursday, for India. They are equipped with the latest modern improvements. They are each of about 2,500 tons burden, 45 feet in breadth, and 285 feet in length, and very handsome. We have never had any ships in this harbor (says the Portsmouth correspondent of the *Times*) which have created such interest as these, for they have been visited by the best sailing and gunnery officers of the navy, and all have expressed their admiration and astonishment at their capacious stowage, airy and ample accommodations, and the unprecedented speed chronicled in their logs.

Their great fame having reached her Majesty through the public journals and the reports of the authorities, and those ships being now within convenient reach of the court at Osborne, her Majesty on Tuesday morning communicated her desire to the naval and military commanders-in-chief at Portsmouth that the embarkation of the troops might not take place until she had inspected them and the ships destined to carry them to their destination. Accordingly, information was given to Captain McDonnell, of the *James Baines* and Captain McKirdy of the *Champion of the Seas*, by the port admiral, of the Queen's intention. Each captain got his ship in order for the inspection, and made every preparation to receive her Majesty. Other troops, not going out in those ships, (the 42nd and 34th regiments,) were also commanded to be in review order in the dockyard by half-past five, as the Queen would inspect them prior to going on board the ships. At a quarter to six the royal steam-yacht *Fairy*, Captain, the Hon. Joseph Denman, having on board the Queen, Prince Consort, the Princess Royal, and Prince Alfred, arrived off Portsmouth harbor. The usual royal salutes were fired by the squadron at Spithead and in harbor, and at six the royal party landed at the new King's stairs of the dockyard, where her Majesty was received by Admiral Sir G. Seymour, K.C.B., Major General the Hon. Y. Scarlett, K.C.B., and staff; Rear Admiral Martin, Col. Foster, R.E., aide-de-camp to the Queen; Lieutenant Colonel Wright, assistant quartermaster general of the southwest district; Capt. Seymour, C.B.; Major Nelson, brigade major; Captain Breton, town major; Dr. Bell, staff surgeon of the district; Lieutenant Hall, director of police; Flag Lieutenants Malcom and Brandreth, etc.

The Royal Marine Artillery kept the ground, and the 42nd and 34th were formed in line from King's stairs to the parade ground leading to the gates, along the face of the storehouses and clock tower, under which a temporary inspecting stage, with standard staff, was placed for the royal circle. The 54th were posted from the Pitchhouse jetty to the Shears jetty, alongside of which the *James Baines, Lady Jocelyn*, and *Champion of the Seas* were moored. Her Majesty, having visited the steamship *Lady Jocelyn*, was conducted by Col. Wright to the *James Baines;* her Majesty was received by Captain McDonnell, and Mr. T. M. Mackay, the owner, at the gangway, and conducted by them over this noble clipper. Her Majesty personally examined the dry and meat provisions supplied for the officers and troops, and expressed her satisfaction at their excellence. She afterwards ascended to the poop and took a view of the great length of deck, thence descended to the troop deck, and walked round it, perhaps the most wonderful and unexampled between-deck her Majesty ever visited, and which appeared to excite her lively surprise. On taking her leave, her Majesty expressed herself much gratified by the visit. She had no idea there were such vessels engaged in the merchant service, and complimented Mr. Mackay and the captain individually on the size and equipments of the *James Baines* and the *Champion of the Seas* generally. We congratulate Messrs. James Baines & Co. on the high honor which has been paid them, and look forward with some hopes to the time when the vessels now employed in the merchant service of Liverpool will be looked upon with more favor by government officials. For the transport or other service no better ships can be employed and the Thames must look to its laurels.

Both ships sailed the same day for Calcutta, and engaged in a race remarkable in its closeness. The *Illustrated London News*, referring to the *James Baines* said:

When met by the *Oneida*, on the 17th of August, on her way to Calcutta with troops, she presented a most magnificent appearance, having in addition to her ordinary canvas, studdingsails, skysails, and moonsail, set and drawing, in all thirty-four sails, a perfect cloud of canvas: the troops all well, and cheering lustily as the vessels passed each other. Her contestant, the *Champion of the Seas*, was not far astern, both vessels making great headway.

The two ships arrived off the mouth of the Hoogly together 101 days from Portsmouth. It was a fight of friendly rivalry all the distance, and one of the greatest ocean races ever recorded in history.

From Calcutta the *Baines* returned to Liverpool with a cargo of Indian produce; arriving there on a Saturday and docking on Sunday, so that the work of discharging her commenced only on Monday. She was almost wholly destroyed by fire April 22, 1858, in the Huskisson Dock; and to this loss is added that caused by the destruction of or damage to very nearly the whole of her cargo. From the first the fire assumed serious aspect, and it was deemed necessary to scuttle the ship. Every possible means was brought into operation for the extinction of the fire, but without success. The time policy under which this ship was insured expiring three days previously and there being no renewal the loss fell directly upon the owners. There was no clue to the origin of the fire beyond the supposition of spontaneous combustion.

Her wreck was converted into the old landing for Atlantic steamer passengers at Liverpool. It is not likely that many of them realized that they were walking over the remains of one of the grandest ships that ever sailed the seas.

CHAPTER XXXII

"The people are gay and colors blow,
The bride but waits for the word to go,
It comes at last and the dog shores fall
'Neath drives and smashes of sledge and maul.

All shout and cry, 'May you take his name,
And on ev'ry sea make known his fame!'
He built her strong, with a counter clean,
The bows sharp out and a dainty beam."

William Brown Meloney
in *The Passing of the Clipper.*

WITH his usual generosity, James Baines showed his sincere appreciation of Donald McKay's services by naming the last of his famous quartette of Boston-built ships *Donald McKay.*

The ship herself was a wonderful combination of beauty and strength. Although not so long nor wide, at the measuring points, as the *Great Republic*, yet she had more spread of floor, was much fuller in the ends, had more cubic capacity, and was the second largest sailing merchant ship in the world. She spread 27 per cent more canvas than the *Great Republic;* the latter, as rebuilt, had only three decks, and her masts and yards were much reduced from their original dimensions.

Where extreme clippers have only space for broken stowage, although her lines were slightly concave below, the *Donald McKay* was decidedly convex above, and hence her great capacity. Her bow preserved its angular form to the rail, and was ornamented with a full figure of a Highlander, "All plaided and plumed in the tartan array" of the ancient McKay clan, or to be precisely Scottish—"MacKay."

This ship measured: length 269 feet, breadth 47, depth 29 feet, deadrise at half-floor 18 inches. She had all the airy beauty of a clipper combined with the stately outline of a ship of war; and, though not sharp, yet her great length, buoyancy and stability, indicated that she could sail very fast, and be an excellent seaboat.

The *Donald McKay* was fitted out with Howes double topsails, a decided improvement over the common rig then in use, and so we describe it below:

The lower topsail yard was trussed to the topmast cap, and instead of slings, was supported from below by a crane upon the heel of the topmast; the lower topsail, therefore, was the size of the closereefed sail of the old rig, and was set entirely by the sheets. The upper topsail set upon the mast above the cap, and had its foot laced to a jackstay upon the top of the yard below, so that no wind could escape between the two topsails. This arrangement of the yards had many advantages. The ship could be reduced to close-reefed topsails at any time, by lowering the upper topsails, which would then lay becalmed before the lower topsails, and the latter, if required to be reefed, could be held without the use of reef-tackles.

In squally weather this rig proved invaluable, for sail could be carried to the last minute, as it could be reduced and reset without a man leaving the deck. Its economy in wear of canvas must also have been very great, for the sails were of manageable size, and had neither buntlines, reeftackles, nor clewlines to chafe them. A ship with this rig was more seaworthy, because she was always considered as under close-reefed topsails, and could be worked with fewer men than a vessel of the same size, having the old rig. It looked rather clumsy in port, and this, we believe, was the principal objection urged against it by those who did not comprehend its advantages at sea. Ships, however, are rigged for service at sea, and not for show in port, that, therefore, which is the most serviceable, is certainly the best.

The *Donald McKay's* masts were nearly upright, her builder contending that they received more support from the rigging than if they raked. In setting up the rigging of a raking mast, it either drags the mast aft, or brings too much strain upon the stays, without giving the mast that side support so essential to make it stand well. Also, in light wind, sails on raking masts flap against the rigging and chafe, while on upright masts they hang clear and almost sleep, giving the ship the full benefit of the breeze. Sails slapping backwards and forwards, often neutralize the effect of a light breeze.

The following facts relating to her principal spars, are interesting. Her lowermasts, topmasts, and bowsprit had, respectively, 30, 33, 18, 5, 5, 3 and 7 tons of pitch pine in them; the lower yards 12, 14½ and 6½; lower topsail yards 8, 8½ and 4; upper topsail yards 4½, 5 and 3 tons;—total 167 tons. She had 3600 feet of chain rigging. The hoops on her fore, main and mizzen masts, were 4 inches wide, by ⅝ths of an inch thick, were 30, 31 and 28 in number, and their weights on each mast 3120, 3210 and 2500 lbs.—total 8830 lbs, hoops on the fore and main yards, 1805 and 2085 lbs.—total 3890 lbs; trusses on the lower yards, 800, 850 and 600 lbs.—total 2250 lbs. Number of yards of canvas in her square sails, 10136; in the fore and aft sails, 4023; studdingsails, 2310; coverings, etc., 286—total, 16755 yards.

Captain Henry Warner, an Englishman resident in East Boston, formerly of the *Sovereign of the Seas*, when under charter to Baines & Co., was placed in command of the *Donald McKay*.

She sailed from Boston February 21, 1855, and made Cape Clear, Ireland, in 12 days, and was 17 days to

SHIP "DONALD McKAY"

From painting by Lars Thorsen, Noank, Conn. Reproduced through courtesy of the artist.

Liverpool. On this, her maiden passage, she ran 421 nautical miles in 24 hours. Donald McKay, himself, went over in the ship and, it is said, he was well satisfied with her.

The *Donald McKay* sailed from Boston in winter and seamen were scarce at the time, so much difficulty was experienced in getting her crew of fifty persons all told. She had only eight seamen on board who were capable of steering or going aloft, the rest were useless "wharf rats," who were obtained from boarding house masters and were bundled on board drunk. More than half of them were seasick most of the passage and unable to come on deck. Fortunately by the newly installed Howes rig she could be reduced to close-reefed topsails, etc., without requiring a man to go aloft; so, although under short sail and poorly manned, Captain Warner made a very rapid Atlantic crossing without accident. Neither he nor his officers took their clothes off to turn in the whole passage.

Upon arrival at Liverpool, the *Donald McKay* was put on the berth for Melbourne, but she did not leave until June and took 81 days to make the run. She left Australia in October and returned home in 86 days. Her average for six consecutive return voyages from Liverpool to Melbourne was 83 days, and only this once did it exceed 85 days. She took upon one occasion 1000 British troops from Portsmouth to Mauritius in 70 days.

After ending her Australian service, she engaged in general trade for a few years, being owned by Thomas J. Harrison of Liverpool, who purchased the *Lightning* and also the schooner *Benin*, 692 tons, constructed for the African trade. This being the only fore-and-after Mr. McKay ever built for foreign account.

Besides the mails, tons of Australian gold dust were carried by these flying clippers, who forged the link between the Mersey and Australia before steam had proved its supremacy upon the sea. Betting was high on their record passages and their arrivals so many days from the Heads were bulletined and acclaimed on 'change in Liverpool and London.

How really fast the McKay clippers were, may be seen by the accompanying authentic performances, which stand, and will stand as world's sailing records, for there never will again be seen on the high seas anything like these splendid white-winged craft; and no finer race of seamen will ever be produced than those who "cracked on" in the clippers.

In 1879 the *Donald McKay* was sold at Bremen and went under the German flag, her hailing port being Bremerhaven. She retained her name and made a number of trips between Bremen and New York, always creating a stir in shipping circles there, not only because of her popularity as one of the last American-built clipper ships afloat, but because of her very large cargo-carrying

INITIAL PASSAGES FROM AMERICA TO ENGLAND

Name	Captain	Voyages	Date	Passage Days	Hours	Best Day	Remarks
Sovereign of the Seas	McKay	N. Y./Liv.	June/July 1853	13	23	344	Dock to Dock
Lightning	Forbes	Bos./Liv.	Feb./March 1854	13	21	436	Boston Light to Dock
James Baines	McDonnell	Bos./Liv.	Sept. 1854	12	9	342	Boston Light to Rock Light
Champion of the Seas	Newlands	Bos./Liv.	June 1854	16	..	..	12 Days to Cape Clear
Donald McKay	Warner	Bos./Liv.	Feb./March 1855	17	..	421	......

INITIAL PASSAGES TO AUSTRALIA

Name	Captain	Voyage	Outward Passage Date	Time Days	Hours	Best Day	Homeward Passage Date	Time Days	Hours	Best Day
Lightning	Forbes	Liv./Melb.	May/July 1854	77	12	348	Aug./Oct. 1854	64	3	412
Champion of the Seas	Newlands	Liv./Melb.	Oct./Dec. 1854	72	..	....	Feb./Apr. 1855	84	..	...
Sovereign of the Seas	Warner	Liv./Melb.	Sept./Nov. 1853	77	..	1275 (4 ds.)	Jan./Apr. 1854	86	..	...
James Baines	McDonnell	Liv./Melb.	Dec. '54/Feb. '55	63	18	423	Mar./May 1855	69	12	...
Donald McKay	Warner	Liv./Melb.	June/Aug. 1855	81	..	.....	Oct./Dec. 1855	86	..	...

The above is taken from "Table of Maiden Voyages of the Principal Early Australian Liners" compiled by F. C. Matthews in *Pacific Marine Review*

capacity and good reliable sailing ability. She eventually went to Madeira as a coal hulk. Her "bones," it is reported, were long lying at Bremerhaven.

> "They hogged her back and ruined her speed,
> And broke her heart with their itching greed,
> So they sold her in the coasting trade
> To carry coals till a grave she made."
>
> W. B. M. in *The Passing of the Clipper.*

TO "DONALD McKAY"

The following verses were written and dedicated to the memory of the distinguished mind that conceived, designed and constructed the *Medium Clipper Ship*, which for beauty, speed and cargo space surpassed all sailing craft built heretofore, succeeding the "Extreme Clipper" as a carrier in world commerce on the seas. These verses are dedicated by an old mariner, who in his youth had the honor—ever since cherished—of meeting Donald McKay personally, as well as Capt. Lauchlan McKay, commander of the famous *Sovereign of the Seas.*

THE "DONALD McKAY"

PIONEER MEDIUM CLIPPER SHIP

Like a huge seabird, with white wings expanded,
Our gallant ship glides o'er the crest of the seas,
Well found, and full manned, and most ably commanded
With a square mile of canvas unfurled to the breeze,

AUSTRALIAN CLIPPERS

With salt sea brine from her martingale dripping
And bowsprit and knighthead well drenched with spray,
She is the nonpareil of all ocean shipping,
The medium clipper ship *Donald McKay*.

With lofty spars and cordage straining,
And huge sails trimmed to the salt sea breeze,
With sparkling spray on the fo'castle head raining,
She sped across the foam-flecked seas.
The *Donald McKay*—then record breaking,
Logged seventeen knots at the very least,
Cargo and passengers, to Australia taking,
Steering a course, about east, south-east.

Those were the days when the Boston-built Clippers,
With huge white wings spread to the breeze,
When Lauchlan McKay and his type of "Skippers"
Won prestige for America upon the seas.
When Donald McKay, the peerless builder,
Added lustre to the Stars and Stripes;
It is enough this old mariner's mind to bewilder,
To think that now they are all "Vanished Types."

Captain Frank Waters,
Sailors' Snug Harbor,
Staten Island, N. Y.

CHAPTER XXXIII

I T is generally conceded that some of Donald McKay's greatest achievements were on the other side of the Atlantic. He had placed Baines' line of Australian clippers ahead of all the world for speed, strength, beauty and completeness, with the *Lightning, Champion of the Seas, James Baines* and *Donald McKay*.

And now we come to two more McKay-built clipper ships that carried the English flag to everlasting fame and glory. In addition to having the famous clipper quartette designed and built for him, James Baines bought from Donald McKay the sister-ships *Commodore Perry* and *Japan* while they were on the stocks in course of construction.

Both vessels were originally designed for McKay's proposed line of European packets, and these dimensions applied to each:

Length 212 feet, breadth 44$\frac{11}{12}$ feet, depth 29 feet. They were somewhat fuller ships than the other four Australian Black Ballers and were designed with a view to carrying large cargoes rather than to attain speed.

Each was fitted with accommodations for cabin and steerage passengers upon nearly the same scale as the Australian clippers previously built by Mr. McKay.

In these ships, their builder had combined two rare elements, buoyancy and stability to a marked degree. They had very little deadrise, but great width of floor. Opposite the main hatchway, across the floor, between the curves of their bilges, each ship was 36 feet wide; and although all spars were aloft and boats stowed on top of their houses, and on gallow frames, each drew only 10½ feet of water on an even keel. They were regarded as "flat bottom" craft and caused much discussion in shipbuilding circles on both sides of the Atlantic. We think their construction entitles Donald McKay to the distinction of being the originator of the "Medium" clipper model, afterwards universally used by American shipbuilders. They were designed before the *Donald McKay*, which many claim was the first in this class.

The furious excitement of the gold rush had died down somewhat and the Liverpool emigrant ships had no longer waiting lists for those desiring passage to Australia. Further, it is to be remembered that homeward cargoes which these clippers brought—wool, hides, tallow, wheat, etc.—were now a commercial factor requiring much consideration, and James Baines readily saw the advantage of using splendid cargo-carrying craft like the *Commodore Perry* and *Japan*.

When she sailed from Boston to Liverpool, the *Commodore Perry's* draught of water in ballast was 12 feet 6 inches, and she distanced every craft that sailed in company with her. She made a record passage from Liverpool to Sydney shortly afterward, thus proving the theory of Mr. McKay that flat bottom was superior to great deadrise.

The *Japan* and *Commodore Perry* remained in Baines' Black Ball fleet for some years. Eventually they passed into other ownership. With a cargo of coal from Newcastle, the *Perry* caught fire and was beached and burned to the water's edge near Bombay, August 27, 1869.

What finally became of the *Japan*, we know not. One story has it that she was condemned and sold at Port Louis on account of her passengers remaining there; bringing about $9,000.00,—which gave each person enough to enable him to leave the island. Originally she had been named *Great Tasmania*. Rebuilt, rechristened by an unknown name, probably she plied the Indian Ocean or China Sea laden with sugar or opium, swarming with centipedes and scorpions, and manned by rice-eating Lascars.

The vicissitudes of life who can explain? After 1860 James Baines' star began to set. He who had been a firm disciple of the sailing ship turned to steam. It must be conceded that his personality covered the fame of the Australian Black Ball Line.

His downfall is undoubtedly traceable to a variety of causes. When he foresaw that the change from sail to steam became inevitable, he made one determined effort to meet the issue by consolidating with Gibbs, Bright & Co., who had started a steamship line to Australia from Liverpool. With Baines' sailing packets and their steamers, the combine kept up a semblance of existence, but from that time he gradually declined as an owner.

The last ships that he had any interest in were the famous *Great Eastern* and the equally famous *Three Brothers*, once the *Vanderbilt*, a paddle steamer of Commodore Vanderbilt's New York to Havre Line, and later a Federal Cruiser. The *Great Eastern* after a rather stormy career was broken up at Liverpool in 1889.

The failure of Barnard's Bank of Liverpool threw James Baines upon his "beam ends." Latterly he had to depend, more or less, on the charity of his friends. He died at Liverpool in 1889, at the age of 66 years.

And so another page of shipping history was turned over, never to be turned back.

PART V

MEDIUM CLIPPER SHIP PERIOD, 1854–1869

CHAPTER XXXIV

AND now the American sailing ship is no longer an extreme clipper, nor is it the bluff freighting craft of 1840, but it is a handsome medium clipper, with towering masts and spars, capable of carrying a good-sized cargo at an excellent rate of speed.

After 1855 there ceased to be a necessity in this country for great speed and size. Too many ships had been built, and a reaction set in, which lasted for many years. A business depression followed the California inflation, which reduced freight rates in the New York-San Francisco trade one-half, and many ships were, in consequence, thrown into the general trade of the world. Owing to the enormously high prices of food agriculture made rapid progress in California, and when it became so productive as to answer her own requirements for food, ships sailing around the Horn experienced much difficulty in procuring cargoes.

America had her "boom" in building wooden ships between say 1843 until 1855; and it was then that steamers began with anything like regularity to enter into competi-

tion with our Boston or New York-built sailing packets.
The struggle for a time was an animated one, and it
should be here noted that about 1854 Donald McKay
made determined efforts to meet the issue, in the establishment of his line of European packets, of which only
two ships, the *Commodore Perry* and *Great Tasmania*,
afterwards named *Japan*, were ever completed. But for
one reason or another the line failed to perpetuate itself
and, as we have previously stated, both vessels were sold
to English shipping interests.

It has often been claimed that Donald McKay originated the large cargo-carrying Medium Clipper. His first
ship that may be so classed is the *Commodore Perry*.
Having fair ends for sailing, with slightly concave lines
below, but almost semicircular above, she deserves the
distinction of being regarded as McKay's pioneer medium
clipper. The *Santa Claus*, a medium clipper launched
September 5th, 1854, was completed ahead of the *Perry*,
which stayed on the stocks many months unfinished.

The reaction of 1857 (helped by extraordinary losses of
vessel property in the winter of 1856 and first part of
1857) dealt a hard blow to American shipping; the
establishment of subsidized lines of British steamers a
still worse one, and what the force of commercial circumstances had nearly consummated the rebel cruisers
completely finished. When American-built ships stood
unrivalled before the world and were in almost universal

demand, the greatest number of bottoms sold to foreigners in any year (1855) aggregated 65,000 tons, whereas the maximum reached in 1864 was 300,865 tons, which aggregated more than our total shipbuilding output for the years 1856, 1857 and 1858.

Our foreign commerce passed into the hands of foreigners, principally because they could afford to build and sail their vessels at much less cost than we could.

Vessels propelled by steam relegated the sailing ship to obscurity, although during the period of which we write, the clipper with favorable winds could outsail nine out of ten ocean steamers then plying the seas. Iron successfully supplanted wood in ship construction, and these factors veritably enthroned Great Britain as Mistress of the Seas. Many contending factors combined to prevent the revival of the American ship upon the high seas, so it followed, naturally, that our shipbuilding industry became paralyzed, and Donald McKay time after time viewed with dismay his once busy shipyard silent and idle. The disruption of that corps of skilled shipbuilding mechanics he had long labored to maintain must have greatly saddened him. He was favorably known far and wide as a generous employer, and in later years he often spoke feelingly of meeting men who had been employed in his East Boston Shipyard, on the Pacific Coast and other far-distant places, following other trades than shipbuilding.

The glorious days when one sharp clipper after another glided down the ways at the McKay shipyard were over, and from the stocks there now came a fleet of staunch ocean carriers. They won no prestige for speed on old Neptune's domain, but they carved their way against powerful odds. The long, long career of Donald McKay's last medium clipper, *Glory of the Seas* (1869–1923, a period of 54 years), is an everlasting tribute to his skill and honesty in the construction of these ships.

CHAPTER XXXV

THE associations connected with her name, drew together many to witness the launch of this ship, and thus another tribute was paid to Daniel Webster, the Defender of the Constitution, showing that though dead, he still lived in the hearts of his countrymen.

Among the spectators were Hon. Edward Everett, Fletcher Webster, Ex-Mayors Seaver and Bigelow, Col. Adams, President of the New England Mutual Insurance Co., Messrs. Crocker, Tappan, Enoch Train, T. J. Shelton, Vernon H. Brown and many others interested in shipping, besides a large number of ladies.

Clipperly in beauty of outline, but packet model in capacity and strength, the *Defender*, designed to stow a large cargo and sail fast, was as fine a vessel of her tonnage as ever cleared the ways at Mr. McKay's yard.

A full figure of Daniel Webster, admirably executed, and painted white, ornamented her bow.

She was owned by Messrs. D. S. Kendall and C. H. P.

Plympton of Boston, and was commanded by Captain Isaac Beauchamp, long and favorably known as one of the best sailors belonging to that city.

Many ladies and gentlemen repaired to Mr. McKay's house after the launch, and there partook of his hospitality. The collation was excellent, and partaken of with a zest heightened by pleasing anticipation of the intellectual treat which was to follow.

Mr. McKay introduced Hon. Edward Everett, as the friend of the lamented Daniel Webster, in whose honor the noble ship was named.

SPEECH OF HON. EDWARD EVERETT
(Phonographic Report for the *Atlas*)

Mr. McKay, Mrs. McKay, Ladies and Gentlemen:— Although I perceive from the manner in which our host has presented me to you on this occasion that something is expected from me, yet I must say I think it is an occasion where anything like a set speech would be not only unusual, but very much out of place. My friend, Mr. Kendall, will bear me witness, if within the sound of my voice, that when he proposed to me the gratification of being present on this occasion nothing was said about speech making; and most certainly I find myself here entirely unprepared for anything formal. By a workman and architect like Mr. McKay, a mere orator must be regarded very much in the light of a land-lubber; and if one should come down here with his tropes and figures on tide water, he would be thought to be playing a superfluous and foolish part. (Laughter.) We look to you, Mr. McKay, for entertainment on this occasion. We have not come here, I

DONALD McKAY'S HOME, 80 WHITE ST. (EAGLE HILL), EAST
BOSTON, MASS.

After launchings at the McKay yard (only a short distance away), the guests
repaired to the "House on the Hill" and enjoyed the hospitality always found there.
This fine old mansion is still standing in an excellent state of preservation, as a
testimonial of its builder's honesty in construction.

"DEFENDER," INDIA WHARF, BOSTON, IN 1857

From a photograph. Kindness of F. B. C. Bradlee. Showing the *Defender* built by Donald McKay in 1855, lying at the end of the wharf on right of picture.

am sure, to have our ears tickled with figures of speech, but to hear the music of your mallet knocking away the last block, and sending off your beautiful vessel to its destined element. (Cheers.)

You were good enough, however, to present me as the friend of that great and good man, whose loss we, as well as the whole country, have so much reason to deplore. For myself, and in behalf of others whom I see around me, I have much pleasure in expressing the great gratification which we have all experienced in witnessing this successful launch. We are indebted to you, sir, and also to the owners, for the tribute you have paid to this great man, who is honored by the name your beautiful vessel bears, as the "Defender of the Constitution." I assure you that it rejoiced my heart to see his well-represented and majestic figure upon the prow, looking down upon the waters as the vessel glided into its appointed element, as he commanded them now that he is gone, as he commanded the hearts of men while living. (Cheers.) Sir, it was a just tribute to his patriotism, to his long and faithful, and must I not add, sir, ill-requited services. He did defend the Constitution, not merely as every good citizen is obliged to defend in duty the government under which he lives, but he defended it because it was the guaranty of inestimable blessings, surrounding us on every side. He defended the Constitution of the United States because he felt, as you and I, and all of us feel, that it is a kind of earthly providence, surrounding us alike while we wake and while we sleep, and assuring us an amount of blessings such as I firmly believe never before were enjoyed by any other people since the creation of this world. (Cheers.)

Sir, there is another reason, another ground on which it was appropriate to give his well earned title—*The Defender*— to this noble vessel. Mr. Webster, among other reasons, extolled the Constitution of the United States because it spread its ægis over the commerce of the country—because it was, in fact, the bulwark of commerce. He knew, as we all know, that commerce was the great civilizer of nations, the parent of liberty, of the arts, of refinement. He knew, sir, from the history of our own country, how the Constitution of the United States had elevated its commerce, from that miserable point of depression which existed before the adoption of the Constitution; when three or four gentlemen, Boston merchants, were obliged to subscribe for the purpose of building and fitting out two or three small vessels, because there was not capital enough in one man's hands to build a vessel alone. He recollected that commerce had joined the States together, and he did not forget what the Constitution had done in establishing this harmonious intercourse between the North, and the South, the East, and the West. Instead of revolutionary legislation, hostile tariffs, and capricious prohibition, which broke up the country and made it into states in reality foreign towards each other, he knew how much the Constitution had done in abolishing that condition of things, and bringing us all into the prosperous intercourse now existing between the several States. But I feel that I am going too far and wearing your patience upon this topic.

This noble ship which has just been launched, will soon spread her canvas abroad to the breeze, but as her Captain— a brave man I am sure he will be, to be entrusted with property of so much value—stands upon her deck and looks towards his home, and sees his native shores melting in the distance, he may be assured that the best wishes of numerous friends, those around us here and others, will follow him over the deep. He may be sure, too, that to no point, however distant, can he carry that vessel where the name and fame of the "Defender of the Constitution" will not have preceded him. (Cheers.) He can enter no port, however distant, where the flag of the Union which his vessel bears will not be a sufficient defence; and sir, I will say one other thing, although you are present, that there is no port, however distant, which he can reach, where a ship built by Donald McKay, will not stand "A, No. 1." (Loud cheers.) Yes, Sir, and if there were any letter coming before A, or any figure standing higher than 1, the vessels of Donald McKay would be indicated by that letter and that figure. (Applause and laughter.)

I was at a little loss, I confess, to comprehend the secret of the great success which has attended our friend and host. [1] Eighty-two ships, I understand, he has built—all vessels such as we have seen today. I do not mean that they were all as large, but they were as well compacted, and looked as splendidly, as they rode into the waves. Eighty-two vessels! No one else, certainly, has done more than our friend to improve the commercial marine of this country, and it has long seemed to me that there was a mystery about it. But since I have been under this roof today I have learned the secret of it—excellent family government, and a good helpmeet to take counsel with and encouragement from. A fair proportion of the credit and praise for this success is, I am sure, due to our amiable and accomplished hostess. (Cheers.) I congratulate also the father of our host, the father of such a son, and the father of such a family. He has, I am told, fourteen sons and daughters, and fifty grand-children. Nine of the latter were born during the last year. I wish to know, my friends, if you do not call that being a good citizen? (Cheers and laughter.)

I am told, ladies and gentlemen, that our friend, Mr. Train, first heard of our host Mr. McKay at Windsor Castle, in England, several years ago, and what he then heard led Mr. Train to place that confidence in him which has never failed to this hour. Now as Windsor Castle, the residence of the British monarchs, was the first place of introduction, may we not, I ask, regard our friend as the "Sea King of the United States." (Cheers.) I will not, however, take up more of

[1] Including six ships Donald McKay built at Newburyport, the total is forty-two.

your time, but as a concluding sentiment I propose that you all drink the health of

Our host, Mr. Donald McKay—A successful voyage to the noble vessel he has launched this morning, and all prosperity to her enterprising owners.

Col. Enoch Train introduced to the company, Hon. Benjamin Seaver, formerly Mayor of Boston, who spoke substantially as follows:

SPEECH OF EX-MAYOR SEAVER

Ladies and Gentlemen,—I am certainly much indebted to you, and to my friend, Col. Train, for this kind remembrance of me. However humble his friends may be, the Colonel always bears them in his remembrance, and cherishes them in his heart; but I shall not by any means be so unwise as to make a speech, after such an entertainment as we have had today. I wish only to bear testimony to one fact, here as everywhere and on all proper occasions, that the city of Boston, the city I love most dearly, is indebted to Donald McKay, I will not say more than, but as much as she is to any other living man within the limits of that city, for its commercial importance, and its reputation for enterprise and energy. By his talents and his enterprise, Donald McKay has been enabled to extend all over the world the fame of the friend graduated at (Mr. Train: "The college of practice"), and I do not care how many languages he may speak, but I am quite sure that he will go down to posterity as one of our great men—as one to whom our city is infinitely indebted for its prosperity and greatness. Having said this much, and these are really the sentiments of my heart, allow me to propose, as a concluding sentiment—

"Health and continued prosperity to our friend Donald McKay."

Colonel Train then addressed the company in substance as follows:

Ladies and Gentlemen.—I must offer a word on this exceedingly pleasant and interesting occasion, before we part. I heard Windsor Castle mentioned just now, by that orator whose voice is a tone from a musical instrument. It carried my mind back certainly eleven or twelve years, to one morning in August when I attended in the chapel at Windsor Castle, where were her Majesty and the royal family. It was that day that I met a gentleman from Newburyport, who told me whom I could employ to build the fastest ships. That commenced my acquaintance with Mr. McKay, and ultimately brought him to East Boston. Since that time he has built, I think, fourteen ships, for me, and from that day to this eighty-two vessels, as we have been told, have furnished ample proof of his enterprise, skill and industry.

Our eloquent friend spoke of the commercial intercourse of the country. Commerce is, as he explained it, the life principle of the nation. It is the Master of Ceremonies which introduces one country to another and sometimes unceremoniously. It knocks at the door of Japan occasionally, and that, too, without regarding the bell pull. (Laughter.) But, as he well knows, and in his remarks he directly and eloquently alluded to the fact, there is a drawer to that bill of commerce. That drawer is labor—is agriculture, without which there could be no commerce. When enough has been produced for home consumption, the merchant steps in and takes the balance and conveys it to the other parts of the world. Am I not right in this? The merchant then is the second rate man. I happen to be one of that kind of second rate men. (Mr. Everett: "It would hardly do for any one else to call you so.") Yes, sir, I am one of those who carry this surplus abroad, and we bring it back again to the country in fancies. But we should always have a little balance to bring back in specie. In order to make us commercially strong, we should have the balance of exchanges in our favor; otherwise we are on the way to ruin. Take a town for example. I recollect going through Georgetown a few years ago, and seeing the grass growing up through the pavements, as I have seen it in towns in the old world. I said then that this town must be growing poor, and that its imports must exceed its exports. This is practical political economy and it is as simple as a farm a mile square. If the farmer imports more than he exports, he is obliged to mortgage his farm, and there is the whole history of political economy brought down to its simplest dimensions. I would, therefore, place agriculture in the first rank, then merchants second, and diplomatists third. (Laughter.)

But, my friend, I am not going to stop yet. In the vicinity of Windsor Castle, in England, I have looked over thousands of acres, which appeared like one vast level forest, with no tree standing above its fellow. It has come into my imagination, since the death of Daniel Webster, that in our American forest the tall trees are also fast becoming extinct. Three or four years ago there were three tremendous tall trees rising up above us all. Now they have all been cut down—Daniel Webster, Henry Clay, John C. Calhoun. (Cheers.) Still, ladies and gentlemen, although the tall trees may have fallen from among us, there is I assure you, a pretty tall sprout still standing—(pointing to Mr. McKay)—there is a man, who if he does not know nothing, gives us abundant assurance that he knows something. (Cheers and laughter.) Ladies and gentlemen, I will say no more about diplomacy.[1]

---

[1] In a case before the Court of Probate, Boston, October 1895, Senator Hoar among other references to Donald McKay's splendid career, said:—

"When the solid men of Boston got together in Faneuil Hall, and Webster or Everett wanted to bring down the house, all that was needed was to allude to Donald McKay, or to speak of the *Defender*, or the *Daniel Webster* or the *Sovereign of the Seas*. If your Honor would like to learn something of what, (if Governor

On September 1, 1855, the *Defender* sailed from Boston bound to San Francisco, arriving there January 14, 1856, making the passage in 135 days.

Captain Beauchamp having been given command of the new ship *Minnehaha*, Captain Robinson took her from New York upon her next two runs to California. On his first passage he reached there March 30, 1857, after 148 days; next year his second passage was made in 150 days time. About six months afterwards, on February 27, 1859, the *Defender* was wrecked on Elizabeth Reef, South Pacific Ocean, while bound from Sydney, New South Wales, with lumber from Puget Sound.

Long and Mr. Morse were not here), I should say were two lost arts, oratory and shipbuilding, which were once the glory of Boston, I wish you would read Mr. Everett's speech at the house of Mr. McKay when the *Defender* was launched in 1855, at East Boston."

CHAPTER XXXVI

"MASTIFF," 1030 TONS,
LAUNCHED FEBRUARY, 1856

THIS was a medium clipper, built by Donald McKay, account of George B. Upton of Boston, for the California and China trade. Her complete outfits were A-1 in every respect, and she enjoyed the distinction of having a library on board costing in the neighborhood of $1,500.

Sailing from Boston March 7, 1856, under command of Captain William O. Johnson, the *Mastiff* reached San Francisco July 19th, 134 days out. She proceeded on her way to China, making one or two stops en route.

It was on one of these long off-shore voyages that she met her fate, by fire, September 15, 1859.

A more graphic picture of an exciting sea episode, than here given by the author of *Two Years Before the Mast*, is difficult to conceive. Remaining in California from the 16th of August to the 10th of September, 1859, Richard Henry Dana then started for the Sandwich Islands, and the following is taken from his disconnected manuscript notes:

September 10. Saturday, 10 A.M. Set sail in the noble clipper ship *Mastiff* for Sandwich Islands. This ship is bound to Hong Kong. Stops at Sandwich Islands to land mail and few passengers, and has one hundred and seventy-five Chinese steerage passengers on board. William O. Johnson, master. His wife on board.

Beats out of harbor exceedingly well. Quick in stays. Last view of San Francisco hills, islands, ports, light-houses, Golden Gate, and its fog and strong northeast winds.

First three days of passage, the coast fogs and cold hold on. Then clear, fair, Pacific Ocean weather, and light winds.

Enjoy highly life in a sailing vessel. So much better than a steamer. No noise, no smell of oil, no tremor, as still as country after city; and the interest in the sails, winds, duties of seamen, etc. Become intimate with Captain Johnson,—seaman by birth, well educated; a library on board which cost some $1,200 or $1,500, and all other things to match—plate cutlery, furniture, provisions, etc.

The ship his home and his idol and chief subject of conversation. He owns one quarter of her, and took her from the stocks; built in Donald McKay's best manner.

Chief mate is Bailey, of New Bedford; second mate, Johnson, of Salem; third mate, a Frenchman, and crew of about twenty men. All newest fashions of rigging.

Interest myself and recall old times by watching working of ship and work on rigging. Songs of sailors. Go below; Chinese burn lamps and smoke. Captain Johnson forbids it.

September 15. Thursday. At about five P.M., quiet afternoon, good breeze, all easy and happy, work going on. Captain Johnson, "Here, Mr. Bailey, fire in the ship!" Startled all; smoke immediately pours up after-ventilator and hatch. Call all hands aft. Rig hose to pump. Mates jump down the hatch aft, in the lazaretto, and smoke pours up in volumes, stifling. Officers spring up and report that between decks all on fire, and, having taken fire in lower hold, Captain Johnson immediately gives up all hopes of saving ship, and stops pumps, and all hands go to work in clearing boats for lowering. "Is there powder on board?" "Yes." Captain Johnson has gone below to get it. Magazine brought up and thrown overboard, and Captain Johnson armed with revolver. Chinese are alarmed, and rush for the boats; beat them back by belaying pins and threats and presenting pistol. Steward shows presence of mind, and stands by captain. Gig is lowered first. Mrs. Johnson comes up, prepared to go in boat.

A British ship has been in sight the last two days, sailing with us. She is several miles astern. Set our ensign union down, and half mast, and back after yards. Captain Johnson asks me to see his wife safely in boat. She goes over side on rope. Chief mate and I help her in. Chinese rush for the boat; beaten back; take in Chinese rower, cabin passengers, and few Chinamen, who rush in. Excellent boat. Second mate takes command; four oars, and I help at one. Pull over two miles and put all safely on board the English ship. Ship *Achilles* bound to Sydney. Calmness of Mrs. Johnson.

Soon two more boats come from the *Mastiff*, each full of Chinamen; one in charge of third mate, other has no officer, so I volunteer to take charge of the boat with a steering oar. Pull for the *Mastiff*. Smoke pouring out, but flames not burst out yet. Put her alongside, and take in Chinese hanging from the sails and ropes and chains. Great noise and attempts to get in, but as they cannot swim are afraid to jump in. Keep boat well off and get her full. Men lie in bottom, and crouch down. Order them aft. Gentle, and ready to obey. Put them all safely on board the *Achilles*. My boat leaks, and keep one hand bailing. Put off again for the *Mastiff*. Five boats now employed—four of *Mastiff*, and one of the *Achilles*, under charge of her second mate. These boats all flying to and fro. Remarkable that with the alarm, and so many (one hundred and seventy-five) ignorant, useless men, not knowing our language, unaccustomed to boats, struggling for life, we should have launched every boat safely, none swamped or stove, and loaded, transported and put on board all—every one—without an accident.

When got alongside last time found all the Chinese had been taken off. Boats now take off baggage of passengers and crew. We had taken none before, Johnson afraid to leave the deck and boats lest Chinese take them.

Steward saves all my luggage, with trifling exceptions, as it was all in my room on deck, and that was to windward. Nothing could be got from below and from lee side. Sailor's house being on deck, save most of their clothes. Captain Johnson saves the specie, $76,000. in gold, in boxes, and chief mate takes it to the *Achilles;* also two chronometers. The Captain saves nothing of his own. Steward saves some trunks for him and for Mrs. Johnson. (Steward's name is Edward Trofater.) Most of luggage in upper house is saved.

Captain Johnson asks me to come on board and have a calm conference to see if anything more can be done. I do so. Very much fatigued by exertions in my boat, especially the steering oar, and head and lungs full of smoke. Captain Johnson says all between decks a mass of fire, and will soon burst out through deck. Cannot get out long boat. Been trying it while we were in boats—too few men, and now of no use. All other boats out, and nothing more can be got from deck. Has been trying to get at the bread, but cannot cut through the deck. (Carpenter of no use, has seemed to lose his powers.) Nor could it have been done, as too near fire. Nothing more can be done. My boat is full of luggage and push off again; put all safely on board. Two boats remain by side of the *Mastiff,* and Captain Johnson, the chief mate, steward, are the last to leave her—not until ordered. Captain last to leave.

Flames burst out through deck at mainmast. Now nearly dark, and flames glow over the ocean. Mrs. Johnson anxious lest her husband stay too long. Two figures on the quarter deck. Now disappear, and the last two boats come off. Captain Johnson comes on board, and the poor, noble *Mastiff* is abandoned. Flames mount the rigging, catch the sails, and all a mass of fire. Main and mizzen mast fall. Foremast stands long, then drops, and only a burning hull.

Captain Hart of the *Achilles*, a generous, frank British sailor, takes Captain Johnson by hand. Now the excitement is over, and his duty done, the magnitude of the loss comes over him, and he says over and over, "My ship *Mastiff!* My ship *Mastiff!* Is it possible she is gone!" Like the mourning of David over Absalom.

The *Achilles* reached Honolulu on the 27th of September, twelve days after the burning of the *Mastiff* and Dana landed there, not failing, as he put off for the shore in a lateen-sail Kanaka boat, to note with pleasure the glorious surf breaking over the coral reefs on each side of the channel through which he sailed.

CHAPTER XXXVII

THIS beautiful three-deck ship of 1695 tons was built by Mr. McKay for Messrs. Kendall & Plympton of Boston, and launched March 22, 1856, at noon, in the presence of quite an assemblage of people. Among those present were America's leading poet, Henry W. Longfellow;* Boston's most popular actress, Mrs. Julia Bennett Barrow, Gen. John S. Tyler; the owners of the ship, and several prominent merchants of the city.

The gentlemen named and some others went on board the ship while she remained on the ways and expressed themselves delighted with the thorough and workman-like manner in which she appeared to be built. She was a full model, designed for carrying rather than speed, and

---

*In Henry W. Longfellow's Journal, under date of March 22, 1856, is this entry:—

"Took the boys over to Donald McKay's shipyard to see the launch of the "Minnehaha". A launch is always beautiful. We went up to Mr. McKay's house, where was a luncheon; and Mrs. Barrow crowned the whole with a recitation."

yet anyone could see by studying her model for a moment that in securing the former requisite the latter had not been sacrificed.

About half-past eleven o'clock, all was ready, the word was given, the hammers of the workmen began their sharp rattle, and she was soon wedged up. The shores were then knocked away, and the ring of the hammers on the iron wedges, splitting away the blocks, was heard. As the workmen approached the last block, under her stem, the spectators seemed to wait in breathless silence to see her start. Still she lingered as if unwilling to leave the shore, and then almost imperceptibly began to move, and gathering force as she proceeded, moved beautifully, gracefully, and majestically into her destined element.

Just as she was clear of the ways, and was bowing her congratulations to friends on shore, Capt. Alden Gifford standing upon her bow, christened her with a bottle of pure Cochituate,—and called her name in a stentorian voice, which was heard even above the cheers of the delighted multitude—*Minnehaha!* And thus the *Minnehaha* was launched.

After the company had spent a few moments in admiring the beautiful proportions of the ship as she sat upon the water, Mr. McKay invited them to his residence on White Street, where a sumptuous repast was prepared by Boston's leading caterer.

After partaking of the collation, Gen. John S. Tyler

ONE OF THE TWIN PARLORS IN THE McKAY HOUSE

In this room Webster, Everett, Garrison, the poet Longfellow, Admiral Farragut, Ericsson, the *Monitor's* inventor-designer, Maury, the great marine scientist, and other famous men were entertained by Donald McKay and his wife.

stated that Mr. McKay had given him the duty of offering a sentiment on the occasion. Speaking, he said,

In the presence of the first shipbuilder, the first poet, and I think I may say the first actress in our country, I feel embarrassed in this position, and I shall detain you only to offer this sentiment:
Success to all three in all their efforts for the glory of New England.

This sentiment was received with great applause.

Professor Longfellow, in response, said—

Mrs. Barrow requests me to return you her thanks for the kind manner in which the allusion to herself has been received. As Mr. McKay is here, he will return thanks for himself.

In proposing to you his health, I would say that it is a singular circumstance that whereas with architecture we generally associate the name of the builder with the structure, yet in this form of architecture, and the most difficult form, because it has to do more than any other with curved lines, we hardly ever hear the name of the builder long associated with the work he builds. I will propose to you——
Success to the *Minnehaha*, and the health of the Builder.

Colonel Adams made a few remarks, illustrating the great public benefit which a man like Mr. McKay was to the community, in furnishing employment and the means of subsistence for many men, etc.

In accordance with the earnest desire of all the company, Mrs. Barrow now repeated several passages from the great poem "Hiawatha." She seemed to have caught inspiration from the occasion, and repeated them with even more than her accustomed charm of interpretation. We have been assured by some persons who were present that the occasion was one of great delight to every one. Mrs. Barrow was then appearing in "Olympia" at the Boston Theatre, after which, by desire and permission of the author, she illustrated, in Indian costume, portions of Longfellow's celebrated Indian poem.

The *Minnehaha* was technically a medium clipper and was considered one of Donald McKay's most beautiful productions. Originally intended for the Australian and China trade, she drifted into other trades.

On her last passage to San Francisco she brought 2330 tons of coal from London. During the voyage she experienced much severe weather, was thrown on her beam ends in a pampero off the Platte, during which the cargo shifted, head was carried away, bulwarks stove and galley and round-house lost. In addition to this the *Minnehaha* was 32 days off the Horn and the run was prolonged further by light winds in the North Pacific to a total of 189 days.

Many fine sailing ships met with disaster in the hazardous fertilizer trade at the mid-Pacific islands, known as the Line Islands, because of their location within and along the equatorial belt, but generally styled the Pacific Guano Islands. The toll taken by their dangerous reefs included the *Minnehaha*. In December, 1867, she was driven ashore at Baker's Island and became a total loss. She was insured for $56,000.00 and at the time of her loss had no cargo aboard. It was reported that she had been sold at New York in May, 1862, for $62,500.00.

CHAPTER XXXVIII

"GLORY OF THE SEAS", 2102, TONS,
LAUNCHED NOVEMBER 1869

"A beautiful and gallant craft
  Broad in the beam, and sloping aft,
  Built for freight and yet for speed."

OUR foreign commerce was rapidly passing into the hands of foreigners and the almost complete stagnation of shipbuilding, principally attributable to enormous taxation, a consequent high cost of material, and high rates of labor, were factors too obviously wrecking American shipping interests, when Donald McKay laid the keel of the *Glory of the Seas*. New ships continued so costly but few merchants would invest in them, so he built and sailed this vessel on his own account, and afterward operated her himself. Reverses and the decline in American commerce had impaired his wealth. Building and owning this—his last merchant ship—in 1869, did much injury to him financially!

She was one of the best built ships ever produced at

McKay's East Boston yard. Viewed broadside on she had all the imposing majesty of a ship of war, combined with the airy buoyancy of a clipper. Registering 2102 tons, with capacity to carry double that amount of freight, she was a splendid model to carry and sail. Her bow had a bold dashy rake, with slightly concave lines below, but convex above, and terminated in a beautifully hand-carved classical female figure, with flowing drapery.

The *Glory*, qualified to class as a medium clipper, was 250 feet long, on the line of the wales, between perpendiculars, and 265 feet from the knight heads to taffrail; breadth of beam 44 feet, 28 feet 6 inches depth of hold. She had three full decks, 8 feet 2 inches height between each deck; had 8½ inches dead rise at half floor, and 7 feet sheer, which was graduated her whole length, with sufficient spring toward the ends to impart an air of lightness and grace to her general outline.

She had a splendid set of spars, finely proportioned and well made. She carried double topsail yards, the yards upon the fore and main masts were alike. A suit of sails contained about 8000 yards of cotton duck, and her rigging was of the best quality. In short, the *Glory*, in all details of construction and equipment, was as nearly perfect as a ship need be. She was christened by Donald McKay's daughter, Frances, now Mrs. Auguste Clavel and residing at Bulle, Switzerland.

Sailing from New York, February 14, 1870, on her

maiden voyage, she reached her anchorage at San Francisco June 13th, 120 days from Sandy Hook. For the first 24 days, or until she reached and crossed the line, she averaged 161 miles per day; also proved a model carrier of freight—not a single package of her 4000 tons of cargo being damaged in the least, either by sea water or sweat.

It is said that on this maiden trip she went out under command of her owner and builder, but our records show she was commanded by Captain Giet, and Captain Chatfield took her from San Francisco to Liverpool. She was then sold to J. Henry Sears of Boston, who put Captain Josiah N. Knowles in command.

She ran between New York and British ports and San Francisco almost exclusively from 1870 to 1885, making one run from New York, October 13, 1873, to the city by the Golden Gate, anchoring there January 18, 1874, in 96 days—the ninth best trip on record.

As soon as the Pacific began to have wheat to export, San Francisco became one of the most profitable ports in the United States for traffic. Wheat raising in California became as profitable as gold mining. American ships at first had this advantage: sailing from the Atlantic states with miscellaneous cargoes of goods, they not only got the freight out, but had a good freight of wheat back to England. Wheat freights from San Francisco to Liverpool rose to a profitable figure, and the *Glory of the Seas,*

The Last Clipper Ship Built by Donald McKay, the *Glory of the Seas*. From a painting by Warren Sheppard. Reproduced through courtesy of Messrs. T. S. & J. D. Negus, New York.

"THE GLORY OF THE SEAS," CAUGHT IN THE TEETH OF A RISING GALE!

Imagine the hustle and activity aboard ship under these circumstances. The wind increases, the ship seems to be just leaping toward the descending sun. First she plunges her bowsprit deep into a white-crowned wave, then drives it skyward as the green water rushes from her deck. Men are aloft, furling the fore uppertopsail, for even with the scant canvas carried the wake is a high flung turmoil of churned-up water.

From painting by Charles Robert Patterson. Reproduced through courtesy of Columbian Rope Co.,

with many another American sailer, entered this grain-carrying trade and made money for her owners.

The one record established by the *Glory* and never equalled was her 35 days passage from San Francisco to Sydney, Australia, in 1875. In some respects it was made under unfavorable conditions. Although wind and weather were favorable, for lack of ballast the ship experienced much difficulty on the voyage.

Captain McLaughlin succeeded to command in 1879 and continued until relieved by Joshua S. Freeman about 1885. Leaving New York December 9, 1879, he arrived at 'Frisco April 5, 1880—118 days out. A long passage, 120 days, California to Ireland, with a cargo of wheat followed. It was on this voyage that the *Glory* narrowly escaped becoming a total wreck. Leaving Queenstown early in the morning, she arrived off the Kish next day during the height of a sudden gale. Finding it impossible, with her 25 feet draft of water, to cross the bar for Dublin, she ran for Kingston where she brought up with both anchors. Here the fury of the gale and the heavy sea running into the harbor ultimately snapped one of her huge cables and, dragging, she took the ground at low water. However, with the assistance of powerful tugs, she was brought into a deep water berth, where she remained till lightened sufficiently to enable her to cross the bar.

After 1885 the *Glory* remained on the Pacific coast,

at one time carrying coal from Puget Sound to San Francisco, with an occasional trip to Alaska, and engaging in various trades. She had some long spells of inactivity, was laid up at San Francisco from December 1882 until February 1885. For a couple of years, from 1904, she was again laid up, this time in Oakland Creek. She could not command paying freights in competition with steel and steam.

Later, deprived of her splendor aloft and her sailing power reduced to a minimum, she was converted into a floating cannery. The idea was to tow the vessel around where the fish were most plentiful and can them right on board. This scheme was not successful, so she was used by the Glacier Fish Company as a floating cold storage plant. It was from their dock in Tacoma that this historic old square-rigger began her last voyage in tow of a puffing tug to be beached and burned for the iron, copper and what other plunder her hull would furnish!

When the hulk of this once famous ship disappeared in flame and smoke, May 13, 1923, on a small pebbly Puget Sound Beach, known as Endolyne, some five miles distant from Seattle, the last sailing vessel Donald McKay built for the American Merchant Marine was lost to the world.

A pretty story to the effect that this almost unique survivor of the splendid line of American sailing vessels, which once queened it over the seas, was to be privately

"GLORY OF THE SEAS"—HER SAD ENDING A FUNERAL PYRE

She was burnt for her copper, on a beach at Endolyne, near Seattle, May 13th, 1923.

purchased with a view to bringing her to Boston for exhibition purposes, after receiving much publicity, was found to be a fake!

So the *Glory of the Seas* ended her days on a funeral pyre!

PART VI

U. S. NAVAL VESSELS BUILT DURING THE CIVIL WAR, AND IN THE YEARS 1874-5—

REPAIRING YACHT "AMERICA"—CLOSING OF SHIPYARD—1861-1875

CHAPTER XXXIX

A ND now we come to Civil War times, with
McKay's shipyard turning out monitors, iron-
clad gunboats and other craft to be used in
defense of the Union. At the commencement of hostili-
ties, Donald McKay unhesitatingly offered his services
to the Government, and placed his entire plant and
equipment at the disposal of U. S. Naval authorities for
repair work. He was often consulted, not only upon
matters appertaining to ship construction, but on various
affairs of importance during that critical period of this
country's history.

Brief historical sketches of the vessels he built for or
sold to the United States Government follow:—

U. S. STEAMER "TREFOIL"

This wooden propeller gunboat was built at East
Boston, in 1864, and purchased by the Government,

February 4, 1865, from Donald McKay for $118,070.00. Her dimensions were—length 145 feet, 7 inches; beam 23 feet, 9 inches; depth 11 feet, 2 inches. Tonnage, 370.

On March 1, 1865, she was placed in commission at the Navy Yard, Boston, Acting Master Charles C. Wells, commanding, and shortly afterwards assigned to West Gulf Squadron.

She saw very little naval service, however, for on August 30, 1865, the *Trefoil* was placed out of commission, and "in ordinary" at Boston Navy Yard, 1866. On May 28, 1867, she was sold to L. Litchfield (Donald McKay's brother-in-law) for $11,500.00.

U. S. STEAMER "YUCCA"

This was a wooden gunboat, screw steamer; length 145 feet, 7 inches; beam 23 feet, 7 inches; depth 11 feet, 3 inches. Tonnage, 373. Built in 1864, she was purchased February 25, 1865, from Mr. McKay for about the same price as the U. S. Steamer *Trefoil*.

Placed in commission April 3, 1865, Acting Master Henry C. Wade commanding, and the following year she served on Gulf Station.

Early during 1868 she was "in ordinary" at Portsmouth, N. H., and on August 26th, she was sold there to R. M. Funkhauser for something like 12½ per cent of Governmental cost.

U. S. STEAMER "NAUSETT"

Like the *Trefoil* and *Yucca* this light draft iron monitor had very little naval service. She was built by contract in 1864, and delivered July 18, 1865.

Went into commission August 10, 1865, at Boston Navy Yard, Acting Master William U. Grozier commanding, and two weeks afterward, August 24, 1865, she was placed out of commission at New York Navy Yard.

From 1866 to 1875, the *Nausett* was laid up at Philadelphia. Her name was changed to *Ætna*, June 15, 1869, but on August 10th, same year, re-named *Nausett.*

John Roach, the well-known shipbuilder, bought her for $3666.00 during August 1875, and afterwards she was broken up.

U. S. STEAMER "ASHUELOT"

Here was an iron side wheel steamer Donald McKay constructed under contract for the U. S. Government early in 1865. Her draft was 8 feet forward, 9 feet aft, and tonnage 1030. He delivered this vessel at the Boston Navy Yard November 30, 1865.

The *Ashuelot* was placed in commission at Boston Navy Yard April 4, 1866, Commander John C. Febiger commanding. She did valiant service in the Asiatic Squadron from 1866 to 1883, proving a striking contrast to contemporary productions by other private builders.

A brief account of her wreck, while under command

of Commander Horace E. Mullan, from Bennett's *Steam Navy of the United States*, so interestingly describes how heroically some of this ship's engineering officers and fire-room crew conducted themselves, that it is here reproduced:—

About four o'clock in the morning of February 18, 1883, the iron double-ender *Ashuelot* proceeding southward along the coast of China from Amoy ran upon one of the Lamock Rocks near Swatow and was so damaged that she sank forty-seven minutes later. Eleven of her people were drowned, the others escaping by the ship's boats to the island, and thence by a Chinese revenue steamer to Swatow. That greater loss of life did not result was due to the circumstance that the ship struck just at the time when both watches were up, one relieving the other. When the order to abandon ship was given, a messenger boy was sent to notify the officer in charge of the engine room, but the boy was so frightened that he jumped overboard without delivering the message and as a result the engine-room force narrowly escaped being left on the sinking ship. Assistant Engineer, J. M. Pickrell, who had stood the mid-watch, was in charge of the watch, assisted by Cadet Engineer W. T. Webster, who had come below to relieve Mr. Pickrell just before the ship struck, these young officers having two watches of firemen below with them.

As the last boat was about to shove off, the executive officer, Lieutenant A. J. Iverson, whose coolness and presence of mind throughout the catastrophe won him high commendation, bethought himself of the engine-room and called down to see if anyone by chance might remain there; to his astonishment he found the two officers and all their men at their posts waiting for orders, although at that time the whole forward part of the vessel was submerged and water was pouring in a broad sheet down into the fire-room over the forward coaming of the fire-room hatch. The boats had all left the ship except the gig, which was still lying under the highly elevated stern and into which Mr. Webster and some of the men dropped from the Jacob's ladder, Mr. Webster being the last person to escape from the ship and at such a late moment that a man behind him on the Jacob's ladder was carried down with the ship and drowned. Mr. Pickrell, being a good swimmer and knowing that the gig could not carry them all, jumped overboard as soon as he came on deck and was picked up in the wreckage over the sunken ship a few minutes later by one of the cutters.

"AFTERMATH"

Later events at Washington, bearing upon the long-delayed reimbursement for the unusual expenses incurred by Donald McKay in constructing these Naval vessels, contrived for many years to make the "McKay Claims" a sort of political football. It certainly was an unbefitting aftermath for his splendid services in the War. From the Boston *Advertiser*, under date of February 10, 1888, the following extract is made:—

Washington, Feb. 10, 1888.—The Donald McKay Claim Bill was passed to the very last stage by the House this afternoon, after a long and rather exciting debate, the great feature of which was Sunset Cox's punishment of Springer for his mulishness and silliness. The McKay bill involves several hundred thousand dollars, claimed by Boston contractors on warships built during the war. The contracts were made after the first encounter of the *Monitor* and *Merrimac,* and the contract was a curious one, so strange as to make one smile. It provided that any changes made necessary by the altered methods of fighting should be made, and if at any time the contractor did not wish to finish his work, the government could go on and do it.

The delays on the work were very great; the changes were more than numerous, amounting in the five cases, of which the McKay's contracts were two, to some millions. Coal and iron went up, in some cases 300 per cent. The price of the cruisers was settled, including the extra material demanded by the change, by a Naval Commission appointed by Lincoln. They allowed for the extra material, but not for the appreciation of material caused by delay. The Supreme Court on appeal later said it was not legal to allow for the raise, but intimated very plainly that the McKays had a very good claim and the case was good in equity. Senator Hoar has had charge of the case from the first. He has written several reports on it and has urged it strongly, having passed it through the Senate before and on this occasion.

On April 4th President Cleveland vetoed the Donald McKay bill. Secretary of the Navy Whitney had earnestly requested him to sign it, but Representative Springer, of Illinois, who for some reason or other conceived an aversion to the bill, informed the President that it was a measure to sign which would offend the "loyal South." As this bill did not propose to take any money out of the Treasury, but only referred the claim of the McKays for a final settlement to the Court of Claims,—President Cleveland's pretext that it was a raid on the Treasury was unfounded upon fact. Donald McKay had been dead eight years and this measure was being pushed through Congress by his youngest brother, Nathaniel.

Years later when the McKay claims were finally adjudicated by the United States Court of Claims, little remained for the widow and children of Donald McKay. Most of the money, some three hundred thousand dollars, not paid out by legal fees, etc., found its way into the pockets of some of that hungry flock of human vultures, the politicians who fattened upon all sorts of claims arising from Civil War exigencies.

Due to "walking the marble halls of Congress" in pursuance of the McKay Claims so many years, Nathaniel McKay became nationally prominent as a lobbyist at Washington and amassed a fortune as the representative of large banking, railroad and other interests at the Capitol. Later on he won renown for the part he played in preventing Grover Cleveland from succeeding himself as President.

CHAPTER XL

R EPRESENTATIONS to Congress of the decaying condition of the navy eventually resulted in a special act, approved February 10, 1873, authorizing the Secretary of the Navy to construct eight vessels of war, the aggregate tonnage of the whole not to exceed eight thousand tons, and the aggregate cost to be not more than $3,200,000. The act specified that four of the vessels, in whole or in part, should be built by the lowest responsible bidders in public competition. Another noteworthy provision of the act was that the ships were to be *"steam vessels of war with auxiliary sail power."* The cost limit was repeated in the naval appropriation passed the following month and placed in the appropriations for the Bureau of Construction, so it did not have to be used to pay for machinery, other appropriations providing for that.

Donald McKay successfully bid for the construction of two vessels under this act—and contracts for the Sloops of War *Adams* and *Essex* were awarded to him.

U. S. SLOOP OF WAR "ADAMS"

(Second of Name)

The *Adams* was classified as a wooden screw steamer by the Navy Department: length, 185 feet; beam, 35 feet; draft, 14 feet, 3 inches; tonnage, 615. She was built at Boston Navy Yard, Mr. McKay working in co-operation with the Government officials. Her machinery was constructed by the Atlantic Works, East Boston, for $163,000. Upon original plans in our possession, this gunboat, or corvette as they called her later, is described as a "Sloop of War." Her keel was laid in 1874.

She was not placed in commission until July 21, 1876. Under Commander John W. Philip, who afterwards distinguished himself in the Spanish-American war when commanding the battleship *Texas*, she was then assigned to the North Atlantic Station. On April 4, 1877, Commander Frederick Rodgers assumed command and she sailed April 19, 1877, for South Atlantic Station. From Montevideo, Uruguay, she went on November 1st, under orders to join South Pacific Station. On November 12th, same year, while off Sarmiento Bank, she received notice of a serious mutiny at Sandy Point, Straits of Magellan; she proceeded there and offered assistance to the Governor, and at his earnest request, remained until security was restored. For the services rendered, Commander Rodgers received the thanks of the authorities of Chili.

On May 10, 1878, the *Adams* sailed with the Samoan Ambassador and his suite on board, and arrived at Apia, Samoan Islands, June 28th.

She continued on the Pacific Station during 1879–1881, cruising off the coast of South America. On February 28, 1880, she established a coaling station at Gulfo, Costa Rica.

On September 11, 1882, she sailed for Alaska, serving in Alaskan waters until August, 1884, to give protection to American persons and property. During her stay in this region, her commanding officers were Commander Edgar C. Merriman, from September, 1882 to September, 1883, and Commander Joseph B. Coghlan from September 1883 to August 1884. In the absence of other officials, they acted as Naval Governors of Alaska. The *Adams* left Sitka on August 10, 1884, and was put out of commission at the Mare Island Navy Yard August 28, 1884.

During 1886 and the early part of 1887, we find her cruising on the West Coast of South America; later sailing for Hawaii, and the Samoan and Tongan Islands. She cruised in the Pacific until December 6, 1888, when she sailed for San Francisco. She was again placed out of commission at the Mare Island Navy Yard on March 25, 1889, but recommissioned April 22nd, and cruised in Samoan waters until placed out of commission July 31, 1890. During 1891, she served in ordinary at Mare Island Navy Yard. On March 29, 1892, she was recommissioned and assigned to the Pacific Station and on May 13th she sailed for Bering Sea for duty in connection with seal fisheries.

The *Adams* was stationed at the Hawaiian Islands during April, May and June, 1893, to protect American interests. She then returned to duty with the Bering Sea Fleet July 2, 1893, remaining until September, 1894, when she returned to Mare Island and went out of commission November 16, 1894. She was recommissioned December 24, 1895, and then cruised in Hawaiian waters until 1896. First five months of 1897 she cruised with apprentices on the Pacific Station. Placed out of commission at Mare Island Navy Yard April 30, 1898. The *Adams* did not serve during Spanish-American War.

She was commissioned at Mare Island Navy Yard October 7, 1898, then placed in Apprentice training service from July 1899, to May 11, 1901, when she once more went out of commission. Being recommissioned August 30, 1892, the *Adams* continued in training service on Pacific Station until 1905.

From 1905 until May 28, 1907, she was Station Ship at Naval Station, Tutuila, Samoan Islands. She left Tutuila for the United States on June 17, 1907, and on November 21st was assigned to the State of Pennsylvania for use as a nautical school ship. She was placed out of commission at League Island Navy Yard, December 31, 1907.

We note she was in use by Public Marine School, Philadelphia, until February 6, 1914, when she was returned to the custody of the Navy Department. Then, on May 1, 1914, she was loaned to the Naval Militia of New Jersey until 1917 when they returned her to the Navy Department. She then became Station Ship in Delaware River, Fourth Naval District, and on January 30, 1919, was assigned to Third Naval District.

On August 5, 1919, for the last time the *Adams* was placed out of commission, and exactly one year afterward she was sold to J. L. Tobin for $15,600.

U. S. SLOOP OF WAR "ESSEX"
(Third of Name)

Under the same form of dual superintendence, as in the case of the *Adams*, this wooden screw steamer, possessing exactly the same dimensions and tonnage, was built by Donald McKay at the Kittery Navy Yard. Her keel was laid in 1874, and the contract for her machinery, at $175,000, was secured by the Atlantic Works.

October 3, 1876, saw the *Essex* placed in commission at the Boston Navy Yard, Commander Winfield S. Schley commanding, being assigned shortly afterward to the North Atlantic Station.

In 1877 she cruised to Liberia and West Coast of Africa. She was attached to South Atlantic Station,

1878–1879, and placed out of commission at Navy Yard, Philadelphia, October 22nd, 1879.

She was recommissioned November 12th, 1881, and attached to Pacific Station through 1882. Joined the Asiatic Squadron January 1st, 1883. On April 16th at Yujai Island, Marshall Group, she took on board Captain S. H. Morrison and part of crew of wrecked vessel *Ranier*. She returned to United States in 1884, via Singapore, East Africa, and Cape of Good Hope. She was placed out of commission at the New York Navy Yard January 15th, 1885. On June 21st, 1886, she was recommissioned and returned to Asiatic Station, via Suez Canal. In October, 1886, she visited Ponapi, East Caroline Group, to investigate reported massacre of Spaniards and afford protection to American missionaries. She left the Asiatic station in January, 1889, returning to United States via Suez Canal, and on May 11th of the same year was placed out of commission at New York Navy Yard.

On April 22nd, 1890, the *Essex* was recommissioned. From June 30th to July 8th, she took part in the Reunion Ceremonies of the Army of Potomac at Portland, Maine, afterwards serving on the South Atlantic Station from October, 1890, to January, 1893. In April, 1893, she was stationed at Annapolis, with cadets on board for instruction. She went out of commission at Navy Yard, Norfolk, Va., June 13th, 1893, but was again recommissioned January 31st, 1894. She served as Apprentice training ship until put out of commission at Navy Yard, Portsmouth, N. H., April 6th, 1898. Like her sister ship, the *Adams*, she took no part in Spanish-American War.

The *Essex* was recommissioned September 29th, 1898, and she continued in training service until placed out of commission at Navy Yard, Portsmouth, N. H., December 5th, 1903.

The Navy Department loaned her to the Naval Militia of Ohio from 1904 to 1916.

In 1917, the *Essex* was placed on duty with the Ninth Naval District at Duluth, Minn. At the present time (1928) this, the last wooden ship Donald McKay built for the U. S. Navy, housed over, engines removed, is used as a Receiving Ship for the U. S. Naval Reserve and the Minnesota Naval Militia.

CHAPTER XLI

THERE is that which should stir the heart of every American, be he sailorman, yachtsman or landlubber, in the thought that completely overhauling the *America*,—winner of the Queen's Cup in 1851, and the most famous yacht in history,—was Donald McKay's last piece of marine work.

Sold in England by Mr. John C. Stevens for $25,000, the *America* was bought there by the Confederate government, brought back to this country, and then sunk in a Southern river to prevent the Federals from capturing her. They raised her, however, and after repairs, made her a tender to a government schoolship in Boston Harbor; but as the raisers claimed prize-money, the Secretary of the Navy ordered her to be sold. General Butler bought her at a Governmental auction sale in 1873. After sailing her for two years, he realized that to compete successfully in yacht races, she needed to be retopped and her masts righted, tiller replaced with a wheel, and many other changes made. In addition to interior alterations, cabins were rearranged and elaborately fitted, so the lawyer-soldier-politician, who was really a picturesque character in American yachting, could enjoy cruising up and down the Atlantic coast, usually with a host of friends, from Maine to Florida.

It was due to General Butler's efforts as a member of Congress, about one year previously, that contracts to build the United States sloops of war, *Adams* and *Essex*, were awarded to Mr. McKay. Furthermore, he was genuinely interested in the *America* which had wrested from England, as his famous clippers had done, much of her vaunted maritime glory, and so he readily consented to "modernize" her. His work, completed about June 15, 1875, undoubtedly had much to do with the good old schooner's success in competing against racing yachts, possessing the advantages of improved conditions in yacht-building, etc., since George Steers designed and built her.

The shipyard that was the birthplace of so many famous packets and other ships, cradled those incomparable American clippers *Flying Cloud* and *Sovereign of the Seas*, and that Ship o' Ships *Great Republic;* the shipyard that launched those remarkable record-breakers, *Lightning* and *James Baines*, and many other wonderful sailing craft—ships such as had never been and never will be again; and that later produced the "Acme of Perfection," combining speed and cargo-carrying in a splendid fleet of medium clippers, terminating in the *Glory of the Seas*, was now forced to close its gates forever.

PART VII
THE VARIED AND LATTER DAY ACTIVITIES
OF DONALD McKAY 1855–1880

CHAPTER XLII

THE VARIED AND LATTER DAY ACTIVITIES OF
DONALD McKAY, 1855–1880

THE man who gave birth to the ideal sailing craft of all time, craft which successfully contended with the elements, outstripped steam for many years, and astonished the world, saw the decline, not of the Roman Empire, but of the Sailing Empire!

In order to employ the force of skilful and well-trained men, which he had gathered together and the capital he had accumulated, his ingenuity was taxed continually, and he passed many sleepless nights. Up to the time of his failure, during a period of about eleven years, it was estimated that Donald McKay had disbursed over seven million dollars in East Boston and had given employment to thousands of mechanics.

PROPOSED OCEAN STEAMSHIP LINE

In May, 1855, intelligence was received at Boston that all the Cunard steamers would probably be required for British Government service in connection with the

Crimean War, and contracts were tendered Donald McKay for building several steamships here which should be taken to Scotland and there furnished with Napier's engines.

McKay, however, with true public spirit and regard for the city and state of his adoption preferred to have a line of ocean steamships constructed here and their engines made by Boston mechanics, deeming them competent to full equality with steamship and engine builders anywhere. The plan was to build "a splendid line of Atlantic steamers rivaling in every respect the Collins Line of New York."

The enterprise received popular approval, because Bostonians felt that there should be an American line of steamships at their port, and a corporation with $1,500,000 capital was proposed.

Later on, according to another account, the Legislature of Massachusetts incorporated Donald McKay, George B. Upton, Enoch Train, Andrew T. Hall and James M. Beebe, under the name of the Boston and European Steamship Company, with a capital of $500,000 "for the purpose of navigating the ocean by steam."

A public meeting was held July 12th, in the interest of this proposed line. A model of a paddle-wheel steamer, to be called the *Cradle of Liberty*, was shown by McKay, which was to cross the Atlantic in six days. Stirring speeches were made by Messrs. George R. Sampson,

E. Hasket Derby and Enoch Train. Mr. Train's remarks are interesting to us, because they show he believed the American sailing packet could compete with steam. "There is a vast difference," he is quoted as saying, "between steam and sailing vessels, and steam would not interfere with his regular business,—the transportation of coarse and weighty commodities, and passengers who could not afford the luxury of a steam passage. He would, instead of opposing the proposed line, lend it the strength of his right arm."

Resolutions were adopted, and a large committee appointed, but this trans-Atlantic steamship project went no further.

INSPECTION OF EUROPEAN SHIPYARDS, TIMBER CONTRACTS WITH BRITISH ADMIRALTY, ETC.

During the fall of 1859, Donald McKay sailed for Europe, not only to complete certain arrangements with the British Admiralty respecting large timber contracts, but, also, to inspect the principal private shipbuilding establishments in England, as well as the Royal Dock Yards. Progressive in his ideas, he was ever on the alert to increase his professional knowledge.

Accompanying are copies of letters from Secretary of the Navy, Isaac Toucey, to Ex-President Franklin Pierce and the latter's reply, also Lieut. M. F. Maury's commendation of Donald McKay.

NAVY DEPARTMENT,
August 18, 1859

DEAR SIR:

I am gratified to have the opportunity of making you acquainted with the bearer hereof, Donald McKay, Esq., of Boston. Mr. McKay holds a prominent position among the Ship builders of America. I can commend him to your courtesy and attention.

I am yours
Most respectfully
I. TOUCEY.

EX-PRESIDENT PIERCE,
Europe.

I desire to add my earnest commendation of Donald McKay, Esq., to that expressed by Secretary Toucey on the preceding page.

FRANKLIN PIERCE.

Sept. 10, 1859.

OBSERVATORY, WASHINGTON,
27th March, 1858.

MY DEAR SIR:

The ships which of late years have won most renown by their performance at sea were built by Mr. McKay, whom I have the pleasure of introducing to you. He is a famous shipbuilder, and for the purpose of seeking an opportunity still further to display his professional skill, he proposes a visit to Denmark. To that end, I commend him to your favorable consideration.

Respectfully &c

M. F. MAURY.

COL. W. DE RAASTOFF
Minister, etc., etc.
Washington.

The following, from Stephen A. Douglas, shows in what esteem he was held by that famous statesman:

UNITED STATES SENATE CHAMBER,
WASHINGTON,
Aug. 12, 1859.

HON. JOSEPH A. WRIGHT,
  United States Minister,
  Vienna.

MY DEAR SIR:

I introduce to you one of our American celebrities, the great ship-builder, Mr. Donald McKay, of Boston. He is a gentleman of immense enterprise, and has built more ships for others and on his own account than any man in our country. His ships are famous at home and abroad; and are unsurpassed, if not unrivalled, anywhere for those qualities which make the "good ship."

It gives me pleasure to ask your friendly attentions in Mr. McKay's behalf.

Very truly yours,

S. A. DOUGLAS.

While in Europe Donald McKay not only visited the governmental and large private shipyards of Great Britain, but he crossed to France, where all facilities were afforded him to examine the dockyards of the Empire. The results of his observations were communicated to the *New York Commercial Bulletin* and other newspapers in America. These letters embodying his views were copied extensively, and when the Civil War broke out, were referred to by naval writers in complimentary terms, with the result that some of his suggestions were adopted by our own Navy Department.

INTENDED BOOK ON NAVAL ARCHITECTURE

Donald McKay, having attained world-wide fame as the designer and the builder of the finest and swiftest ships that ever sailed the seas, was prevailed upon while in England to undertake the task of authorship. A Prospectus, dated at London, November 8th, 1859, describes his intended work on the Theory and Practice of Naval Architecture, particularly illustrating American Ship Building,—as forming a large octavo volume of about 300 pages, accompanied by an Atlas of 60 large folio plates, giving the lines of clippers and other vessels built by McKay and the details of construction of the same ships.

How regrettable that this project was not carried through to successful completion cannot be expressed in words. The construction drawings and sail plans of so few of his famous ships are obtainable today. Anyone familiar with the difficulty of securing accurate information or data anent the progress of naval architecture, etc., during a period, say, from 1840 to 1859, can certainly appreciate the inestimable loss sustained by Donald McKay's inability to make such a notable contribution to the world's scientific literature. A whole revolution in naval architecture was crowded into those years.

LATTER DAY ACTIVITIES OF DONALD McKAY

INACTIVITY PRIOR TO CIVIL WAR—ONLY ONE SHIP AND FOUR SMALL SCHOONERS CONSTRUCTED FROM 1856–1861

No craft was launched during the year 1857. McKay's shipyard was closed to wait for better times. Sometime in 1858, work was commenced upon the *Alhambra*, a ship of 1097 tons, which remained on the stocks over a year.

Cape Cod fishermen have always prided themselves on owning the swiftest vessels of their class afloat, and a few of the most ambitious in the trade now engaged Mr. McKay to build them a model craft that would outsail any vessel belonging to either Cape Ann or Cape Cod. He turned them over to his son, Cornelius W.,—who produced a schooner which soon was without a rival,—the *R. R. Higgins*. Soon thereafter three other fore-and-afters were constructed, the *Benjamin S. Wright, Mary B. Dyer*, and *H. & R. Atwood*. This, approximately 1600 tons, constituted the entire output of McKay's yard, for a period of over four years until the breaking out of the Rebellion.

THE AMERICAN NAVY AS COMPARED WITH THE ENGLISH

Because many erroneous statements with regard to American Naval affairs, underrating our power to sustain a maritime war, were broadcasted by the English press during the early part of the Civil War, Donald McKay, who was in London, sent a long communication to the *Star and Dial*.

After expertly setting forth details of our Naval force which could be made immediately available for defense, he continues:

Our facilities for building a new fleet are greater than those of any of the European Naval powers—I think, even I may safely assert, greater than those of them all combined.

I will only add [he writes] that I have made this statement not out of any feeling of animosity, but merely with the intention to show the resources of our country for carrying on a defensive war, and to show that whatever the ultimate result of an aggressive war on our country may be—hundreds of ships, and many thousands of men will have gone to the bottom of the sea before its end will have arrived.

In conclusion he states—

I feel perfectly satisfied that in all well-meaning minds of both countries the most friendly mutual feelings exist, and that a war between England and the United States would prove the greatest calamity the world has ever seen, and therefore let us pray that the leading men of both countries may not rule their actions by a misplaced pride, jealousy or animosity, but by the true interests of their respective countries.

Such powerful arguments so clearly advanced by a man internationally known and respected for his knowledge and skill in maritime, as well as naval matters, bore considerable weight and must have affected the judgment of thinking men who read them.

Appreciating that the public had been quite restless in relation to the movements of the Navy Department on account of the depredations of the *Alabama, Florida* and *Georgia,* Donald McKay thought it his duty to all concerned to publish, in the *Boston Daily Advertiser,* an interesting letter which created favorable widespread comment.

Although it especially referred to a disagreement between the contractors of some recently constructed light-draft monitors and the Navy Department, his open and frank way of imparting information to the public concerning certain Naval matters in which they were much interested, could be advantageously followed by government officials and others today.

After giving John Ericsson much praise and credit for having saved the nation a great humiliation through his construction of the first monitor, he advances, as his opinion, that two monitors which Ericsson designed and superintended, *Dictator* and *Puritan,* excelled in material, workmanship and invulnerability anything he saw in England or France.

Referring to the *Alabama* as inferior in speed and fighting qualities to our sloops of war, the truth of which bold assertion was clearly demonstrated in a then recent naval combat, he goes on to say—

The difficulty has not been a want of vessels of the right class to destroy these British rovers, but to obtain sight of them. It is my deliberate opinion that almost any of our sloops could easily overhaul the *Florida* in a twelve hours' run, and bring her to action. The accounts of her great speed, from the size of the vessel and her well-known motive power, are entirely incorrect.

Concluding he advises:

If the public will only exercise a little patience they will find that the Navy Department has not neglected its duty in this hour of our national struggle.

Patriotic sentiments, well expressed and made at a time when much criticism was levelled at our Naval Authorities.

DONALD McKAY UPON OUR NAVY

(*From a communication to the Boston Daily Advertiser, October 28, 1864.*)

When war was upon us, my interest in our navy was increased. I knew the strength of England and France upon the ocean, and the respect which that strength elicited from other nations, and with many others, came frequently before the public with a view of developing our own naval resources. I saw that these powers respected nothing but force, and knew that if we desired to prevent foreign intervention in our affairs, our navy must be largely increased. I refer to these facts at this time to show that when I write of naval affairs, I write of that which I know, and which will stand the test of fair-minded scrutiny. Nothing is easier than to find fault, and I can state from my own experience, that with all my care, *I never yet built a vessel that came up to my own ideal; I saw something in each ship which I desired to improve.* But I should have felt unfairly dealt with if my ships had been judged by their blemishes rather than by their merits. The same rule I propose to apply to a review of our naval affairs, that those interested may see what has been done, and is still doing by the Navy Department.

From the foregoing one readily appreciates that the shipbuilder, whose wonderful sailing vessels had in times of peace carried the Stars and Stripes to glory, was not found wanting when America needed him.

In addition to expressing his very practical and valuable suggestions with regard to American Naval affairs, Donald McKay reviewed exhaustively, in this most interesting letter, much of what the Naval Administration had done, during the war, giving facts and data which proved his thorough knowledge and active interest. Erroneous statements were being put forth by many writers who evidently knew but little whereof they affirmed, and Mr. McKay's efforts to strengthen the arms of government created favorable comment upon all sides.

BUILDING IRON SHIPS, WOODEN STEAMERS, MARINE ENGINES, ETC.

From an attractive booklet entitled "The Ship *Great Republic* and Donald McKay, her Builder," written by Francis B. C. Bradlee, the following is extracted:

In order to keep up to the times, Donald McKay, during the Civil War, changed his yard over so that he could now build iron ships, marine engines, etc. The business was carried on under the name of McKay & Aldus.

At this yard were built in 1863, as a private venture, several wooden steamers—the *General Hooker, General N. P. Banks, Charles W. Thomas, Edward Everett,* etc. Most of them were sold to the Government for transports during the war. In 1866, Mr. McKay also constructed the wooden paddle-wheel steamers *La Portena* for the Cuban coastwise traffic, and *La Orientale* for Capt. William T. Savory of Salem, who intended to run her in the South American coast trade. On one occasion Mr. McKay told the late Capt. George L. Norton that the building of the latter vessel had caused him nearly as much trouble as all his others put together.

Besides shipbuilding Mr. McKay took up the construction of railroad locomotives; turning out several fine ones for the old Eastern R.R. (of Massachusetts), the Fitchburg R.R., the Little Rock & Fort Smith R.R. of Arkansas and others.

In 1869 the firm of McKay & Aldus sold out to the Atlantic Works, a corporation still in active business, and Nathaniel McKay, who was the leading spirit in the enterprise, embarked in another business venture.

PROTESTING AGAINST THE NON-REGISTRATION OF AMERICAN SHIPS BY BRITISH LLOYDS

From a newspaper article written by Mr. McKay in 1872, we quote:—

The favorite English motto "Fair Play," is lost sight of by British Lloyds, when we ask them to give vessels of our flag the rating we merit and should receive on their books. You cannot find our vessels entered in the register of British Lloyds, nor a classification given one; but as soon as the flag is changed and she becomes a British vessel, they will then rate and class her on the same principle as their own build, and we are called selfish and unjust for not opening free trade to the English flag. In all the marine registers of our country, every vessel entering our ports is rated and classed, no matter under what flag she may be, and it is done without prejudice, too, for in many cases I know the rating of foreign ships here in the United States has been higher than would be given our own ships built of the same material.

THE DECLINE OF AMERICAN SHIPPING INTERESTS

It was a lamentable fact that the enormous duties levied upon nearly all articles required for the construction and equipment of a ship, did as much to diminish our shipping for some years after the Civil war, as the depredations of the "British pirates" during the contest. Our foreign commerce was rapidly passing into the hands of foreigners, because they could build and sail their vessels at nearly 50 per cent. less cost. As a partial relief, Donald McKay endeavored to get a bill through Congress permitting a drawback of duties on materials entering into the construction of new vessels, but without success.

Congressional committees officially visited shipyards for the purpose of reporting to Congress the condition of American shipbuilding, but after waiting years for the fulfillment of promises, many shipbuilders gave up the hope of ever laying any more keels and either retired or engaged in some other trade.

CHAPTER XLIII

DONALD McKAY'S LAST DAYS

REPAIRING the yacht *America* practically ended Donald McKay's labors in his shipyard at East Boston. At this time he was suffering from incipient consumption, but, possessed of the Demon of unrest, work he must, and breaking up his home in East Boston, he moved his family to a farm in the town of Hamilton, Mass. Here his heart of oak, so long and gallantly unyielding, at last was stilled. For with his characteristic energy, the residue of his waning vitality, which, if properly nursed, would have spared him to his family for years to come, was hurled against the barren hills of Eastern Massachusetts, in futile endeavors to wring from such sterile soil an independent living for himself and family. Time and again he dropped down in his fields from sheer exhaustion, and on the 20th of September, 1880, like a child sleeping, brain and body worn out, this once tireless, energetic "Son of Toil" crossed the Dark River. He lies buried in Oak Hill Cemetery, Newburyport—a sweet spot on the top of a small hill, fascinating in its enchanting stillness to a wearied soul; where

"AFTER LIFE'S FITFUL FEVER HE SLEEPS WELL."

APPENDIX I

LIST OF SHIPS BUILT BY DONALD McKAY

CONSTRUCTED AT NEWBURYPORT, MASS.

NOTE:—In 1841, Currier & McKay built at Newburyport the Barque *Mary Broughton*, of 323 tons register, for N. Broughton of Marblehead, Mass., and in 1842 the Ship *Ashburton*, 449 tons, account of C. Hill and Captain A. Plumer of Boston; then the *Courier*, of 380 tons, was built and designed by Donald McKay. Shortly after, having formed another partnership, McKay & Pickett, he built the ships *St. George, John R. Skiddy* and *Joshua Bates*, upon completion of which Mr. McKay removed to East Boston, late in the year 1844.

NAME AND CLASS OF VESSEL	DATE LAUNCHED OR WHEN BUILT	BY WHOM AND WHERE OWNED	CAPTAIN'S NAME	TRADE IN	REGISTERED TONNAGE	REMARKS
Courier Trading Ship	1842	Andrew Foster & Son, N. Y.	W. Wolfe	Rio Janeiro, etc.	380	
St. George Packet Ship	1843	Capt. W. Ferris, D. Ogden & Co., N. Y.	W. Ferris	N. Y.–Liverpool packet (Red Cross Line of N. Y.)	845	Burned in the Chops of the English Channel.
John R. Skiddy Packet Ship	1844	Wm. and Francis Skiddy, N. Y.	W. Skiddy	N. Y.–Liverpool packet	930	
Joshua Bates Packet Ship	1844	Enoch Train & Co., Boston (1st White Diamond Line Packet built by D. McKay)	Murdoch	Boston–Liverpool packet	620	In 1862 passed under British flag being bought by Francis E. Beaver of Melbourne. Two years later she was sold to Lowe Kong Meng, a Melbourne merchant, and in 1872 to Wm. Henry Bean of Adelaide. Within a few months she was condemned at Mauritius.

CONSTRUCTED AT EAST BOSTON, MASS.

Name and Class of Vessel	Date Launched or When Built	By Whom and Where Owned	Captain's Name	Trade In	Registered Tonnage	Remarks
Washington Irving Packet Ship	Sept. 15 1845	E. Train and Capt. E. Caldwell, Boston	Eben Caldwell	Boston–Liverpool packet (Train's White Diamond Line)	751	Went under English flag about 1852 and was engaged in Australian trade afterwards.
Anglo Saxon Packet Ship	Sept. 5 1846	E. Train & Co., Boston	Joseph R. Gordon	Boston–Liverpool packet (Train's White Diamond Line)	894	Lost off Cape Sable late in winter of 1846—upon her second voyage.
New World Packet Ship	Sept. 9 1846	Wm. and Francis Skiddy, Grinnell Minturn & Co., N. Y.	Wm. Skiddy	N. Y.–Liverpool packet (Swallow Tail Line)	1404	Sold to Austrians in 1882 and named *Rudolph Kaiser.* Last voyage reported was from Ship Island (Pascagoula) to Sunderland—Sept., 1884.
Ocean Monarch Packet Ship	July 1847	E. Train and Robt. G. Shaw, Boston.	Murdoch	Boston–Liverpool packet (Train's White Diamond Line)	1301	Destroyed by fire August 24, 1848, off Orm's Head (near Liverpool). Large loss of life and ship completely destroyed.
A. Z. Packet Ship	Oct. 1847	August Zerega & Co., N. Y.	Ricker	"Z" Line–N. Y.–Liverpool packets	700	
Anglo American Packet Ship	Feb. 1848	Enoch Train & Co., Boston	Albert H. Brown	Boston–Liverpool packet	704	Name changed to *Arrogant* and sold under English flag about 1852; afterwards engaged in Australian trade, etc.

Name / Type	Date	Builder / Owner	Captain	Trade	Tons	Remarks
Jenny Lind Packet Ship	May 1848	Fairbanks & Wheeler, Boston	Bragdon	Cotton carrier—sailed in Southern trade etc. Also trans-Atlantic to Liverpool	533	
L. Z. Packet Ship	Dec. 1848	August Zerega & Co., N. Y.	Ricker	N. Y.–Liverpool packet	897	
Plymouth Rock Packet Ship	Feb. 13 1849	Geo. B. Upton and Capt. Eben Caldwell, Boston	Eben Caldwell	Boston–Liverpool packet (Chartered by Train & Co.)	960	
Helicon Barque	May 1849	Capt. Adams, Boston		East Indies	400	
Reindeer Ship	June 1849	Geo. B. Upton, Boston		California and Manila	800	**Wrecked on a coral reef, coast of Zambales, P. I., Feb. 12, 1859.**
Parliament Packet Ship	Dec. 1849	Geo. B. Upton and John Forbes, Boston	Albert H. Brown	Boston–Liverpool packet (Train's White Diamond Line)	998	
Moses Wheeler Trading Ship	March 1850	J. P. Wheeler and Capt. King, Boston	James B. King	General business	900	
Sultana Barque	June 1850	Edw. Lamb & Bro., Boston	Watson	Smyrna	400	
Cornelius Grinnell Packet Ship	June 1850	Grinnell, Minturn & Co., N. Y.	Wm. Howland	N. Y.–Liverpool packet (Swallow Tail Line)	1118	Sold at auction March 27, 1883, and changed into a coal barge. Had been seriously damaged by fire a short time previously.

367

Name and Class of Vessel	Date Launched or When Built	By Whom and Where Owned	Captain's Name	Trade In	Registered Tonnage	Remarks
Antarctic Packet Ship	Sept. 1850	Zerega & Co., N. Y.	Ricker	N. Y.–Liverpool packet ("Z" Line)	1115	
Daniel Webster Packet Ship	Oct. 1850	Enoch Train & Co., Boston	Wm. H. Howard	Boston–Liverpool packet	1187	Sold in 1856 after Train & Co. failed. Name changed to *Hygeia*.
Stag Hound Extreme Clipper	Dec. 7 1850	Geo. B. Upton and Sampson & Tappan, Boston	Josiah Richardson	California	1534	Lost by fire off Pernambuco, Oct. 12, 1861. All hands saved.
Flying Cloud Extreme Clipper	April 15 1851	Train & Co., sold to Grinnell, Minturn & Co., N. Y.	Josiah P. Creesy	California and East Indies	1782	Stranded on coast of New Brunswick, near St. John; after, when being repaired, took fire and was so badly damaged only fit for scrap heap.
Staffordshire Extreme Clipper	June 17 1851	Capt. Albert H. Brown & E. Train & Co., Boston	Albert H. Brown	Joined Train's White Diamond Fleet, Boston and Liverpool—then California and East Indies, returning to Atlantic route.	1817	Lost off Cape Sable when bound from Liverpool to Boston, under command of Capt. Josiah Richardson, Dec. 29, 1854. The Captain and 169 persons were lost.
North America Extreme Clipper	Sept. 1851	Nickerson & Co., Boston	A. Dunbar	Liverpool	1464	

Ship	Date	Builder	Commander	Trade	Tonnage	Remarks
Flying Fish Extreme Clipper	Sept. 1851	Sampson & Tappan, George B. Upton, Boston	Edward Nickels	California, China and East Indies	1505	Wrecked when coming out of Foo Chow, Nov., 1858, and then abandoned. Subsequently floated and rebuilt, name changed and sailed some years, finally foundering in China Sea.
Sovereign of the Seas Extreme Clipper	July 1852	Donald McKay, Boston	Lauchlan McKay	California and East Indies, also Australian and General Trade	2421	Ran on Pyramid Shoal, Straits of Malacca, in 1859 and proved a total loss. Was on a voyage from Hamburg for China.
Westward Ho! Extreme Clipper	Sept. 14 1852	Sampson & Tappan, Boston	Johnson, later Hussey	California and China Trade	1650	Purchased by the Peruvians and went under their flag — name unchanged. Caught fire on Feb. 27, 1864, when in harbor of Callao and burned until she sank at her moorings.
Bald Eagle Extreme Clipper	Nov. 25 1852	Geo. B. Upton, Boston	Philip Dumaresq	California and China Trade	1704	Sailed from Hong Kong, October, 1861, for San Francisco and never heard from since.
Empress of the Seas Extreme Clipper	Jan. 14 1853	Donald McKay (Sold on stocks to Wm. Wilson & Sons, Baltimore)	M. E. Putnam	California and China Trade	2200	Burned at Queenscliff (Port Phillip) on December 19, 1861.
Star of Empire Extreme Clipper	April 1853	E. Train, and T. Hall and Benj. Bangs, Boston	Albert H. Brown	Boston and Liverpool Packet Service	2050	Lost at sea.

CONSTRUCTED AT EAST BOSTON, MASS.

NAME AND CLASS OF VESSEL	DATE LAUNCHED OR WHEN BUILT	BY WHOM AND WHERE OWNED	CAPTAIN'S NAME	TRADE IN	REGISTERED TONNAGE	REMARKS
Chariot of Fame Extreme Clipper	May 1853	E. Train, Benj. Bangs, A. T. Hall and Capt. Knowles, Boston	Allen H. Knowles	Boston and Liverpool Packet Service. Afterwards Australian Trade	2050	Abandoned when bound from Chincha Islands to Cork in January, 1876.
Great Republic Clipper	Sept. 4 1853	Donald McKay When rebuilt—A. A. Low & Brother	Lauchlan McKay, Joseph Limeburner	General Trade (When launched) Troopship Crimean War, California Trade (As rebuilt)	4555 3357	Burned when loaded ready for sea lying at Peck Slip, New York. Paid for by the insurance. Bought by A. A. Low and Capt. Nathaniel B. Palmer. Rebuilt and ran in various trades until 1869, when she changed flag and was renamed *Denmark*. Caught in hurricane en route Rio de Janeiro to St. John, N. B., to be repaired. All hands reached land in safety.
Romance of the Seas Extreme Clipper	Nov. 15 1853	Geo. B. Upton, Boston	Philip Dumaresq	California and China	1782	Loaded at Hong Kong for San Francisco sailing Dec. 31, 1862. In April, 1863, was officially posted as missing. Thirty-five lives were lost with her.
Lightning Clipper	Jan. 3 1854	Jas. Baines & Co., Liverpool, Eng.	James Nicoll Forbes	Liverpool to Australia. Used as troop ship Sepoy Mutiny	2083	Burned at Geelong (Melbourne Harbor) Oct. 31, 1869, and scuttled at her anchorage.

Name	Date	Builder / Owner	Captain	Trade	Tonnage	Remarks
Champion of the Seas Clipper	April 19 1854	Jas. Baines & Co., Liverpool, Eng.	Alexander Newlands	Liverpool to Australia. Used as troop ship, Sepoy Mutiny	2447	Foundered off Cape Horn in 1876. All hands taken off by British barque *Windsor*.
James Baines Clipper	July 25 1854	Jas. Baines & Co., Liverpool, Eng.	Charles McDonnell	Liverpool to Australia. Used as troop ship, Sepoy Mutiny	2525	Before being unloaded was almost wholly destroyed by fire April 22, 1858, in the Huskisson Dock, Liverpool.
Blanche Moore Extreme Clipper	1854	Charles Moore & Co., Liverpool, Eng.		Liverpool and East Indies	1787	
Santa Claus Medium Clipper	Sept. 5 1854	Joseph Nickerson & Co., Boston	Bailey Foster	California and Calcutta	1256	Abandoned in a sinking condition, August 9, 1863, attempting to make St. Thomas.
Benin Schooner	1854	Thos. Harrison & Co., Liverpool, Eng.		African trade	692	
Commodore Perry Medium Clipper	1854	D. McKay. Sold to Jas. Baines & Co., Liverpool, Eng.		Australia and Liverpool	1964	Caught fire and was beached and burned to the water's edge near Bombay, August 27, 1869.
Japan Medium Clipper	1854	D. McKay. Sold to Jas. Baines & Co., Liverpool, Eng.		Australia and Liverpool	1964	Originally named *Great Tasmania*.
Donald McKay Clipper	Jan. 1855	Jas. Baines & Co., Liverpool, Eng.	Henry Warner	Liverpool to Australia. Also General Trade	2594	Went to Madeira as a coal hulk; afterwards, it is reported, taken to Bremerhaven and left there.

CONSTRUCTED AT EAST BOSTON, MASS.

NAME AND CLASS OF VESSEL	DATE LAUNCHED OR WHEN BUILT	BY WHOM AND WHERE OWNED	CAPTAIN'S NAME	TRADE IN	REGISTERED TONNAGE	REMARKS
Zephyr Medium Clipper	1855				1184	Final destination unknown.
Defender Medium Clipper	July 28 1855	Kendall & Plympton, Boston	Isaac Beauchamp	General business	1413	Wrecked Feb. 27, 1859, on Elizabeth Reef, South Pacific Ocean, while bound for Sydney with lumber from Puget Sound.
Henry Hill Clipper Barque	1856	Chas. Brown and Capt. Watson, Boston			568	
Mastiff Medium Clipper	Feb. 1856	Geo. B. Upton, Boston	Wm. O. Johnson	California and China	1030	Burned at sea Sept. 15, 1859, when five days out from San Francisco, bound for Hong Kong.
Minnehaha Medium Clipper	Mar. 22 1856	Kendall & Plympton, Boston	Isaac Beauchamp	California and China	1695	Driven ashore at Baker's Island, December, 1867, and became a total loss.
Amos Lawrence Medium Clipper	1856	Emmons & Parsons, Boston		California and East Indies	1396	
Abbott Lawrence Medium Clipper	1856	Geo. B. Upton and Capt. E. Caldwell, Boston		California and East Indies	1497	

372

Name / Type	Date	Builder / Owner	Commander	Service	Tonnage	Remarks
Baltic Medium Clipper	Oct. 1856	A. Zerega & Co., N.Y.	Ricker	N. Y.–Liverpool Packet Service "Z" Line	1372	
Adriatic Medium Clipper	1856	A. Zerega & Co., N.Y.	Ricker	N. Y.–Liverpool Packet Service "Z" Line	1327	
Alhambra Medium Clipper	1858–9	Wm. Thwing & Co., Boston			1097	
R. R. Higgins Schooner	1858	Ulrich, Mayo & Co., Boston		Fishing—Cape Cod Fleet		
Benj. S. Wright Schooner	1859	B. S. Wright & Co.,		Fishing—Cape Cod Fleet	107	
Mary B. Dyer Schooner	1860	M. F. Dyer & Co. Boston		Fishing—Cape Cod Fleet		
H. & R. Atwood Schooner	1860	Hicks Atwood, Boston		Fishing—Cape Cod Fleet		
General Putnam Ship	1861–2	Geo. B. Upton, Boston				
Trefoil Screw Propeller Wooden	1864–5	U. S. Government	Acting Master Charles C. Wells	Naval Service	370	Placed out of commission in 1866. Sold by Government May 28, 1867.
Yucca Screw Propeller Wooden	1864–5	U. S. Government	Acting Master Henry C. Wade	Naval Service	373	Sold in 1868 for something like 12½% Governmental cost.
Nausett Iron Clad Monitor	1864–5	U. S. Government	Acting Master Wm. W. Grozier	Naval Service		Sold August, 1875, and broken up.

CONSTRUCTED AT EAST BOSTON, MASS.

Name and Class of Vessel	Date Launched or When Built	By Whom and Where Owned	Captain's Name	Trade In	Regis-tered Tonnage	Remarks
Ashuelot Iron side-wheel double ender	1864–5	U. S. Government	Comm. John C. Febiger	Naval Service	1030	Ran upon one of the Lamock Rocks near Swatow, China, and quickly sank, February 18, 1883
Geo. B. Upton Screw Propeller Wooden	June 25 1866	D. McKay, Boston		Atlantic Coastwise	604	
Theodore D. Wagner Wooden Screw Propeller	July 4 1866	Boston & Charleston Line of Steam Packets		Atlantic Coastwise	607	
North Star Brig	1867				410	
Helen Morris Medium Clipper	1867–8				1285	
Sovereign of the Seas II Full Model Ship	1868	Lawrence, Giles & Co., N. Y.			1502	Went under German flag and renamed *Elvira;* again known by name she first bore; converted into coal barge and lost off Barnegat in 1910.
Glory of the Seas Medium Clipper	Nov. 1869	D. McKay, Boston	Giet	California, etc.	2102	Burned for her metal May 13, 1923, at Puget Sound.

374

Frank Atwood Schooner	1869	F. M. Dyer & Co., Cornelius W. McKay and others, Boston		West Indies	107	Carried "Bill" Tweed from New York to Cuba—then sold. [5]
Adams Sloop of War	1874-5	U. S. Government	Comm. John W. Philip	Naval Service	615	Placed out of commission for last time August 5, 1919, and exactly one year afterward she was sold to private party.
Essex Sloop of War Completed by Contract	1874-5	U. S. Government	Comm. Winfield S. Schley	Naval Service		Now used (1928) as Receiving Ship for U. S. Naval Reserve at Duluth, Minn.
America Schooner Yacht (Retopped and righted up her masts, etc.)	1875	Gen. Benjamin F. Butler, Boston		Private Yacht		Stationed (1928) at U. S. Naval Academy, Annapolis, Md.

APPENDIX II

RECORD PASSAGES OF DONALD McKAY'S CALIFORNIA (EXTREME) CLIPPER SHIPS

Ship	Port of Departure	Date of Sailing	Arrival at San Francisco	Days	Hours
Stag Hound	New York	Feb. 1, 1851	May 26, 1851	108	
		(Total time at sea—113 Days)			
Flying Cloud	"	June 3, 1851	Aug. 31, 1851	89	21
Flying Fish	Boston	Nov. 11, 1851	Feb. 17, 1852	98	
Staffordshire	"	May 3, 1852	Aug. 13, 1852	101	
Sovereign of the Seas	New York	Aug. 4, 1852	Nov. 15, 1852	103	
Flying Fish	"	Nov. 1, 1852	Feb. 1, 1853	92	
Westward Ho!	Boston	Oct. 20, 1852	Feb. 1, 1853	103	
Bald Eagle	New York	Dec. 26, 1852	Apr. 11, 1853	107	
Flying Cloud	"	Apr. 28, 1853	Aug. 12, 1853	105	
Westward Ho!	"	Nov. 14, 1853	Feb. 28, 1854	106	
Romance of the Seas	Boston	Dec. 16, 1853	Mar. 23, 1854	96	
Flying Cloud	New York	Jan. 22, 1854	Apr. 20, 1854	89	8
Stag Hound	"	Apr. 27, 1854	Aug. 14, 1854	110	
Flying Fish	Boston	Sept. 22, 1854	Jan. 10, 1855	109	
Westward Ho!	"	Jan. 17, 1855	Apr. 24, 1855	100	
Flying Cloud	New York	Feb. 18, 1855	June 6, 1855	108	
Flying Fish	Boston	Sept. 13, 1855	Dec. 27, 1855	105	

(No Record Passages from Atlantic ports to San Francisco were made by McKay clippers during year 1856.)

Westward Ho!	New York	Dec. 16, 1856	Mar. 26, 1857	100	
Flying Fish	Boston	June 24, 1857	Oct. 2, 1857	100	

When California freights fell down, the Cape Horn Clipper Fleet became disorganized, so the McKay flyers, with many other American ships, sailed in other trades or went under foreign flags.

It is interesting to note that during the many years that have elapsed since the close of the Civil War, although a large number of sailing ships have been built for the California trade, Donald McKay's last production, the *Glory of the Seas*, holds the record for having sailed from New York to San Francisco in 94 days.

INDEX

INDEX

Ashuelot—Continued
tions by other private builders, 333;
heroism displayed by this ship's
engineering officers and fire-room
crew, when she was wrecked off
China coast, 334–335.

Atlantic, Record Passages Across
(Table of fast runs by McKay
clippers on maiden voyage), 289.

Atlantic Works, East Boston, con-
structs machinery U. S. Sloop of
War *Adams*, 339; secures contract
for machinery of the *Essex*, 342;
buys out the firm of McKay & Aldus,
362.

Atwood, Hicks, Boston, shipowner, 373.

Australia, Table of Outward and
Homeward Passages made by Mc-
Kay clippers, 289.

Australian Clippers (*See* James Baines
& Co., also *Lightning; James Baines;
Champion of the Seas; Donald
McKay; Japan; Commodore Perry*,
259–295; *Chariot of Fame*, 103).

A. Z., New York–Liverpool "Z" Line
packet ship, 56, 366; her cargo list
upon arrival New York in May,
1850, from Liverpool, 58.

Baines, James, Liverpool merchant
and shipowner, 259–261, 292; his
bust serves as a figurehead on ship
James Baines, 277; names the last
of his famous clipper ship quartette
Donald McKay, 283; buys *Japan* and
Commodore Perry while on stocks,
292; vicissitudes attending his later
life, his last ships, *Great Eastern* and
Three Brothers, dies at Liverpool, 295.

Baines, James, & Co., Liverpool merchants and owners, Black Ball Line
of Australian packets, 192, 195–196,
257–258, 259–262, 267–268, 281,
292; challenging advertisement for
Sovereign of the Seas—offering to
return freight money if she did not
make faster passage than any steamer
on the same berth, 197; *Empress of
the Sea* arrives at Melbourne 66½
days out from Liverpool, is burnt at
Queenscliff, 228; vessels for, 370–371.

Bald Eagle, California clipper ship,
369; a charming description of a
"deep sea lady," 210–211; her
metallic lifeboat, 214; account of
maiden voyage, with extract from
log, 216–217; Capt. Dumaresq is

succeeded by Capt. Caldwell, 217;
engages in Chinese coolie trade, 218;
leaves Swatow for Callao with about
700 coolies aboard and duly delivers
her human cargo, 220; a fake yarn
about this clipper ship's destruction,
220–223; contradiction and a true
account of her voyages, &c., 224;
sails from Hong Kong and is never
heard from afterwards, 224; sailing
records to California, 376.

Baltic (medium clipper), New York–
Liverpool "Z" Line packet ship,
60, 373.

Baltimore clippers, 113–114.

Bangs, Benjamin, Boston merchant-
owner, 104, 369, 370.

Barrow, Mrs. Julia Bennett, Boston's
most popular actress, 317; repeats
passages from Longfellow's poem
Hiawatha at luncheon after *Minnehaha* launching, 319–320.

Beauchamp, Capt. Isaac, *Defender*,
304, 372; is given command of
Minnehaha, 311.

Behm, Capt. C. F. W., *Stag Hound*,
130, 132.

Bell, Jacob, builder, New York, 9, 14;
sends Donald McKay to Wiscasset;
15.

Benin, schooner, African trade, 371,
owned by Thomas J. Harrison of
Liverpool; only fore-and-after Mr.
McKay ever built for foreign account, 288.

Benjamin S. Wright, schooner built
for Cape Cod fleet, 357, 373.

Bennett's *Steam Navy of the United
States*, extract from, describing
wreck U. S. steamer *Ashuelot* off
coast of China, 334–335.

Black Ball Line of Liverpool, Baine's
Australian packets, 259–262, 277;
taking same name as trans-Atlantic
Packet Line, causes confusion, 195–
196; list of initial (record) passages
to Australia, 289. (*See* James Baines
& Co., also James Baines.)

Black Ball Line (New York packets),
33, 85; request James Baines to
adopt another name for his Australian line, without success, 195–196.

Blanche Moore, medium clipper ship,
East India trade, 371.

Boole, Albenia Martha (first wife of
Donald McKay), 10–11; dies after
a brief illness, 56–57.

378

INDEX

China Clippers, 253; poetry eulogizing the sailing clipper, 254.

Clark, Capt. Arthur H. (author, *Clipper Ship Era*), 272, 278.

Clavel, Mrs. Auguste (Donald McKay's daughter, Frances), christens *Glory of the Seas*, 323.

Cleveland, Grover, President of the United States, vetoes "Donald McKay bill," 336.

Clipper Ship Races—Lieut. Maury's account of the famous Cape Horn race of 1853, 170–176; great discussion and heavy bets made on ships as to relative speed and length of passage, 165; wholesome sporting interest and enthusiasm in these long off-shore ocean sailing matches, 252; betting high on Australian clippers; their record passages and arrivals bulletined and acclaimed on 'change in Liverpool and London, 288.

Coghlan, Commander Joseph B., U. S. N., *Adams*, 340.

Commodore Perry, clipper ship built for McKay's trans-Atlantic line—afterwards used in Australian trade, 260, 300, 371; details of construction, 292–293; makes record passage Liverpool to Sydney; remains in Baines' Black Ball fleet some years; catches fire, is beached and burned, 294.

Complete Account of Voyage Around the Horn (*Sovereign of the Seas*, in 1852), 184–188.

Condry, Dennis, owner of *Delia Walker*, 15; recommends Donald McKay to Enoch Train, 20.

Congressional Committees visit shipyards to study shipbuilding conditions, 363.

Coolie Traffic, description of, 208–209; cruel conditions attending transportation of men, their behavior aboard ship, 219–220; terrible tragedy of the *Bald Eagle*, 220–223.

Cope's Line of Philadelphia (Philadelphia to Liverpool packets), 33.

Cornelius Grinnell, New York–Liverpool, Swallow Tail Line, packet ship, 367; details of construction, 86–87; a poem depicting Christmas dinner on board, 88; is damaged by fire, 89; changed to a coal barge and "towed" where she formerly sailed in majestic splendor, 90.

Cotton Ships, peculiarities of construction and their cargoes, 73–75; manned by a hard class of men called "hoosiers," to whom is attributed chantey singing aboard American ships, 75–76.

Courier, Rio Trader (first ship built and designed by Donald McKay), 365; her success as fast sailer, brings her builder before maritime public, 17.

Cradle of Liberty, Donald McKay's model of paddle-wheel steamer which was to cross Atlantic in six days, 352.

Creesy, Capt. Josiah P., *Flying Cloud*, 144–145, 149, 154, 368; extracts from log, 146–148; reads his own "Obituary" while at sea, 149–150; letter accompanying presentation of silver service set from Walter R. Jones, President of the Board of Underwriters, 158–159; his reply, 160; with Mrs. Creesy, he decides to give up the sea, 160–161; appointed Commander in U. S. Navy, assigned to clipper *Ino*, 162; subsequently commands *Archer* and makes two voyages to China in her; died at Salem in 1871, 162–163.

Crews—American Clipper Ships during California Clipper period composed largely of vagabonds, 149; description of a scene in a New York shipping office during the California gold rush, 167-170; when there was a scarcity of seamen at San Francisco, 206.

Crews—American Packet Ships, 59; to "hoosiers" who worked and sailed aboard "Cotton Ships," is attributed chantey singing, 75–76.

Crews—Australian Sailing Ships, carried large crews, England to Australia, but managed with smaller ones returning home; behavior of the crews in Australian waters, 267.

Crew of *Donald McKay* on maiden voyage across the Atlantic in winter of 1855; only eight seamen capable of steering or going aloft; advantageousness of Howes' double topsail rig proved, 287.

Cunard Line, 68–69.

Cunard, Samuel, 36–37 (*See* Cunard Line).

Currier & McKay (William Currier), 16; firm dissolved, 18.

INDEX

Essex, Sloop of War, U. S. Navy, 375; placed in commission, assigned to North, then South Atlantic Station, 342; at Yujai Island, Marshall Group, takes on board captain and part of crew wrecked *Ranier;* placed out of commission and re-commissioned, until she serves as apprentice training ship, 344; continues in training service, then Navy Department loans her to Naval Militia of Ohio; on duty now (1928) at Duluth as a Receiving Ship for Minnesota Naval Militia, 344.

Everett, Hon. Edward, 303; his speech at launching of *Defender*, 304–308; Senator Hoar's reference to this speech (October, 1895), 310–311.

Fairbanks & Wheeler, Boston merchant-owners, 72, 367.

Farragut, Admiral D. G., U. S. Navy, 200.

Farrell, James A., vi–vii; his Foreword, ix.

Fastest Day's Run, *Lightning* on maiden voyage, Boston to Liverpool, logs 436 miles in 24 hours, 264–265; makes second greatest day's run of 430 miles, when bound to Australia, 267–268.

Febiger, Comm. John C., U. S. N., *Ashuelot*, 333, 374.

Ferris, Capt. W., *St. George*, 365.

Figureheads, original from *Great Republic* now at Stonington, Conn., 243; interesting description of *Champion of the Seas* full figure of a sailor, 272; "all plaided and plumed in the tartan array" of the ancient MacKay clan was the figurehead of the *Donald McKay*, 283.

Florida, Confederate cruiser during Civil War, 359–360.

Flying Cloud, California clipper ship, 368; origin of name, 141–142; details of construction, 142–143; commences eventful voyages to Frisco, 145, 154; mutiny breaks out, 146; establishes record to California, 147–148, 155; condition of ship when she came home, 150–151; commencement of second voyage New York to Frisco, 151; races against *N. B. Palmer* and reaches Golden Gate three weeks ahead, the *Palmer* having had mutiny on board, 152;

beats clipper *Hornet* next voyage to California, 153; Maury's description of long ocean race with *Archer*, 155–156; deciding merits of *Flying Cloud* and *Andrew Jackson*, 156–157; another quick run to China, 157; interesting description of voyage from China, in letter of presentation silver service to Capt. Creesy, 158–159; last passage to San Francisco under Creesy's command, 160; American sailing epoch at end and Civil War commences, 161; is purchased by James Baines, of Liverpool, and goes under British flag, afterwards sold to Smith Edwards; ends her career by fire, 162; sailing records to California, 376.

Flying Fish, California clipper ship, 154, 204, 206, 369, noted as a contestant in ocean races; race against *Swordfish*, 165–167; shipping conditions and how sailors were shipped during California rush, 167–170; the great clipper race of 1853, 170–171; Maury's account of this sailing contest around the Horn, 171–175; engaged in various trades and finally founders in the China Sea, 176; sailing records to California, 376.

Forbes, Capt. James Nicol, ("Bully" Forbes), *Lightning*, 261, 263–264, 370; originates the slogan "Melbourne or Hell in sixty days!" 265; leaves *Lightning* to take charge of ill-fated *Schomberg*, 266.

Forbes, John, Boston shipowner, 367.

Forbes, Capt. Robert B., originator salt pickling, 46; first American to apply double topsail yards to large shipping; his rig used on *Great Republic*, 233; as President Sailors' Snug Harbor writes Donald McKay; $1,000.00 is collected from visitors to the *Great Republic*, which enabled starting this home, etc., 234–236; when *Republic* rebuilt Howes' rig is substituted, 242; advantage of double topsails, 250–251.

Foster, Andrew & Son, owners New York, 365; give Donald McKay his first commission to design and build a ship, 16.

Foster, Capt. Bailey, *Santa Claus*, 371.

Frank Atwood, schooner, 375.

Freeman, Capt. Joshua S., *Glory of the Seas*, 325.

INDEX

Funch & Meincke, N. Y. ship brokers, 192.

General Putnam, Boston ship, 373.
George B. Upton, wooden screw propeller, 374.
George Washington, N. Y. packet ship, 47.
Gibbs, Bright & Co., Liverpool, Steamship owners of Australian line, and James Baines form partnership, 295.
Giet, Capt., Glory of the Seas, 324, 374.
Gifford, Capt. Alden, 42, 231, 318.
Giles, Lawrence & Co., New York shipowners, 374.
Glory of the Seas, medium clipper ship, 302, 374; unsatisfactory conditions prevailing when this ship's keel was laid, 322; her construction, christened by Donald McKay's daughter, 323; sails on maiden voyage with her owner-builder aboard; arrives at San Francisco in 120 days with 4000 tons of cargo A1 condition; makes run of 96 days, New York to Frisco, sails between New York and British ports, and San Francisco, 1870 to 1885, carrying wheat from California, 324-325; establishes record of 35 days, San Francisco to Sydney; narrowly escapes being wrecked off Irish Coast, 325; remains on Pacific Coast, engaging in various trades, with long spells of inactivity, later is converted into a floating cannery, afterwards a cold storage plant; begins her last voyage, is beached and burned for "junk," 326; a pretty story she was to be privately purchased, proves a fake, so her days ended on a funeral pyre, 326-327; sailing records to California, 376.
Gordon, Capt. Joseph R., Anglo Saxon, 366.
Great Eastern, 295.
Great Republic, clipper ship, 370; when launched the largest and finest ship in the world, what she symbolized, 229; her launch a great occasion, 230-231; details of construction and description some of her outfits, masts and rigging, 230-233; list of materials used in her construction, number of days' work on her hull, etc., 234; Lauchlan McKay placed in command, 234; correspondence carried on by Capt. R. B. Forbes and Donald McKay to aid "Sailors'

Snug Harbor" in Boston, 235-236; tried under topsails and courses, and sailed so fast she dragged towboat astern of her, 236; her destruction by fire; considered a national calamity, 236-237; as rebuilt, McKay's creation is no more, 239; incontrovertible facts to prove erroneous repeated statements that Donald McKay never recovered from loss of this ship, 240-241; surrendered to insurance companies, bought by Capt. N. B. Palmer for A. A. Low & Brothers, then rebuilt, 242; sails first voyage Capt. Limeburner, commanding, and makes quick run to Liverpool, 243; chartered by French Government to carry troops in Crimean War, 243-244; returns to New York; successfully employed in California trade makes record California run of the year; crosses the line in fastest sailing time on record, 245; laid up at Falkland Islands for repairs about six months; travels between New York and San Francisco next two years; sails to England with one of California's first grain exportations, 246; seized at New York as rebel property shortly after Civil War breaks out; chartered to U. S. Government; afterwards resuming her part in California trade; sold after being laid up about two years in New York, 247; again sold and renamed Denmark, changed flag; employed in East Indian Trade; caught in a hurricane off Bermuda springs a leak and sinks, 247-248; comparison with Baines' ship Donald McKay, 283-284.
Great Tasmania, later named Japan, 294 (See Japan).
Griffiths, John W., 11-12, 83, 113, 114, 259.
Grinnell, Minturn & Co., New York, owners, 33, 50, 85, 89, 142, 145, 182, 183, 366, 367; operated Atlantic packet ships until early eighties, 54; print Flying Cloud log in gold, on white silk, 151.
Grinnell, Moses H., New York Merchant-owner, 50, 85, 142, 182-183.
Grozier, Wm. W., U. S. N., Acting Master Nausett, 333, 373.

H. & R. Atwood, schooner built for Cape Cod Fleet, 357, 373.

383

INDEX

Hall, Andrew T., Boston merchant-owner, 104; one of incorporators, Boston & European S. S. Co., 352.

Hammond, Capt. William H., *New World*, 54.

Harrison, Thomas J., Liverpool merchant and shipowner, 371; purchased *Lightning*, also schooner *Benin* (the only fore-and-after Mr. McKay ever built for foreign account), owner *Donald McKay*, 288.

Helen Morris, medium clipper ship, 374.

Helicon, Barque, East India trade, 367.

Henry Clay, New York packet, 47.

Henry Hill, clipper barque, 372.

Heritage of Tyre (*See* Wm. Brown Meloney).

Heroic conduct of Engineering Officers and Fire-room crew at sinking of U. S. Navy steamer *Ashuelot*, 334–335.

Hersey, E. L., superintendent McKay's shipyard, 231.

Honolulu to New York, record passage, (*Sovereign of the the Seas*), 190.

Hornet, California clipper, races *Flying Cloud*, 153.

Hour's Run, best ever made by a sailing vessel, 278.

Howard, Capt. Wm. H., *Daniel Webster*, 92, 106, 368; rescues passengers ship *Unicorn*, 93–95.

Howes, Capt. Frederick (of Brewster, Mass.), inventor of double topsail rig; rebuilt *Great Republic* carried this rig, 242; application of double topsails the greatest improvement of all time in the rig of square-rigged vessels, 250–251; *Donald McKay* fitted out with Howes' double topsails; description of same, 284–285; practical example of the advantages of Howes' rig shown on this ship's maiden voyage in winter of 1855, 287.

Howland & Aspinwall, New York merchants and shipowners, 114.

Howland, Capt. Wm., *Cornelius Grinnell*, 86, 367.

Hussey, Capt. Samuel B., *Stag Hound*, *Westward Ho*, 133, 207, 369.

Illustrated London News, extract from, referring to the *James Baines* sailing with *Champion of the Seas* to Calcutta, 281–282.

Inactivity Prior to Civil War, only one ship and four schooners (small) constructed at McKay's shipyard from 1856 to 1861, 357.

Indenture, Articles of, Apprentice's. (*See* Apprenticeship in New York Shipyard.)

Independence, New York packet ship, 47.

Initial Passages from America to England, 289; to Australia, 289.

Ino, California clipper ship in U. S. Navy, commanded by Capt. Josiah Creesy, 162.

Inspection of European Shipyards, by Donald McKay, 353.

Iron Ships, Wooden Steamers, Marine Engines, etc., their construction at the McKay Shipyard, 361–362.

Iverson, Lieut. A. J., U. S. Navy, *Ashuelot*, his conduct at sinking wins him high commendation, 334.

James Baines, clipper ship built for Australian service, 274–275, 292, 371; carries troops to India during Sepoy Rebellion, 268; sails from Boston Light to Rock Light, Liverpool, in 12 days, 6 hours, 276; adjudged the finest ship Liverpool has ever seen; makes an unbeatable record, Liverpool to Melbourne, 277; comes home in 69½ days; completes best round-the-world-passage 133 days, 278; makes highest rate of speed, ever made by a sailing vessel, 278; during Sepoy mutiny she goes to Portsmouth for troops, sailing for Calcutta, with *Champion of the Seas*, 278–279; graphic account of a visit of Queen Victoria and her royal party to two American-built clippers, designed by Donald McKay, 279–281; one of the greatest ocean races in history, remarkable in its closeness, 281–282; returning to Liverpool from Calcutta is destroyed by fire and scuttled; her wreck converted into old landing for Atlantic steamer passengers at Liverpool, 282; record passages, 289.

Japan, clipper ship built for McKay's trans-Atlantic Line, afterwards engaged in the Australian trade, 260, 300, 371; details of construction, 292–293; remained in Baines' Black Ball fleet some years; her end one of the mysteries of the Indian Ocean or China Sea, 294.

INDEX

Jenny Lind, cotton-carrier—also trans-Atlantic packet ship, 367; engaged making "Triangular Run," 72–73.

John Gilpin, California clipper ship, races against *Flying Fish*, 170; details of contest, 171–176.

John R. Skiddy, New York packet ship, 20, 365.

Johnson, Capt. Wm. O., *Westward Ho, Mastiff*, 205, 369, 372; placed in command of *Mastiff*, 312.

Joseph Walker, clipper ship, destroyed by fire with *Great Republic*, 237.

Joshua Bates (Pioneer of Train's White Diamond Line of Boston-Liverpool packets), 20–21, 37, 365.

Kamiesch (Crimean War), an enthusiastic American traveller writes about the many American clippers there, 244.

Kendall, D. S., Boston merchant and shipowner, 303, 317.

Kendall & Plympton, Boston merchants and Shipowners, 303, 304, 317, 372.

King, Capt. James B., *Moses Wheeler*, 367.

Knight, Capt. Hale, *New World*, 54.

Knowles, Capt. Allen H., *Chariot of Fame*, 104, 106, 107–108, 370.

Knowles, Capt. Josiah N., *Glory of the Seas*, 324.

Lamb, Edward, & Bro., 367.

Lardner, Dr. (and other scientists), arguments against Atlantic steam navigation, 33–35.

Launchings, 47–49; launch of *Great Republic*, 230–231. (See "List of Ships Built by Donald McKay" for Date of Launch or When Built.)

Lightning, clipper ship built for the Australian trade, 259, 261, 271, 274–275, 292, 370; first of the historic American clipper quartette that rendered wonderful service to England and Australia, 261–262; details of construction, 262–263; starts on her wonderful (maiden) voyage, 263–264; makes a phenomenal run, 436 miles in 24 hours, 264–265; on her first passage to Melbourne, Capt. Forbes originates slogan "Melbourne or Hell in sixty days!" On return trip hangs up record of 64 days, 265; Capt. Enright takes command, 266; "wood butchers of Liver-

pool" fill in concave lines of this clipper's bow, letter from Donald McKay, upon the subject, also stating he saw "436 nautical miles" recorded in her log, 266–267; three years later, when bound to Australia, she makes second greatest day's run, 430 miles, 267–268; engaged in transportation of English troops to India during Sepoy Rebellion, sustains her reputation for speed; comparative table of passages, 268–269; returns to Australian trade; is sold to Thomas Harrison of Liverpool; meets her end by fire, 269. Note: 59 years afterwards her remains are uncovered by a dredge, 270; record passages, 289.

Limeburner, Capt. Joseph, *Great Republic*, 243, 244, 245, 246, 247, 370.

Litchfield, L. (Donald McKay's brother-in-law), purchases U. S. steamer *Trefoil*, 332.

Litchfield, Mary Cressy, marries Donald McKay, 193. (*See also* Mary Cressy McKay).

Lloyds, British, newspaper article written by Mr. McKay protesting against the non-registration of American ships, 362–363.

Logs: Extracts from *Flying Cloud's*, 146–148, 154–155; Grinnell, Minturn & Co., print *Flying Cloud* log in gold upon white silk, 151; Lieut. Maury's account of Famous Clipper Race of 1853, between *Flying Fish, John Gilpin, Wild Pigeon* and *Trade Wind*, taken from the logs of these ships 171–175; extract taken from log of *Bald Eagle* on maiden voyage, 216–217; what happened on board *Lightning* when she logged 436 miles, as a day's run, 264–265; her builder states in a letter that he saw "436 nautical miles" recorded, 266–267; Australian clipper *James Baines* records "ship going 21 knots with main skysail set," the highest rate of speed ever made by a sailing vessel, 278.

Longfellow, Henry W., American Poet, a frequent visitor at McKay yard, extract from his poem, 143; *The Building of the Ship* perhaps an incentive for *Great Republic*, 230; present at launch of *Minnehaha*; extract from his "journal" referring

INDEX

Pickrell, J. M.—*Continued*
and is picked up in wreckage over sunken vessel, 335.

Pierce, Franklin, ex-president of the United States, 353; letter recommending Donald McKay. 354.

Plummer, Captain, *Washington Irving*, 106.

Plymouth Rock, Boston-Liverpool, White Diamond Line, packet ship, 367; some construction details, 79; bears an enterprising Yankee expedition into the South Seas, 80–81.

Plympton, C. H. P., Boston merchant and shipowner, 303–304, 317 (*also* Kendall and Plympton).

Proposed Ocean Steamship Line, 351-353.

Protest against the non-registration of American ships by British Lloyds, 362–363.

Putnam, Captain G. W., *Daniel Webster*, 96.

Putnam, Captain M. E., *Empress of the Seas*, 227, 369; *Wild Pigeon* race with *Flying Fish*, etc., 170–175.

Races Around the Horn, 165–167; details of famous clipper ship match of 1853, 171–176.

Rainbow, first extreme clipper, 12, 83, 113, 114.

Record passages, list of, California Clippers, 376; Australian clippers, 289 (*See also* various ships).

Red Star Line of New York-Liverpool packets (Robert Kermit & Co., owners), 33.

Reindeer, general trading vessel, 83, 367; her non-American crew, 84.

Richardson, Capt. Josiah, *Stag Hound* and *Staffordshire*, 99, 100, 106, 368; placed in command of *Stag Hound*, 126; letter to Sampson & Tappan, 128; leaves to take *Staffordshire*, 129; loses his life with his ship, 102.

Ricker, Captain, *A. Z.*, *L. Z.*, *Baltic* and *Adriatic* ("Z" Line, New York Liverpool packets), 60, 366, 367, 368, 373.

Roach, John, New York shipbuilder (afterwards moved to Chester, Pa.), purchases U. S. Steamer *Nausett*, 333.

Rodgers, Commander Frederick, U. S. Navy, *Adams*, 339.

Romance of the Seas, California clipper ship, 370; details of construction,

249–250; sails from Boston to San Francisco in 96 days, 251; races with *David Brown*, beats her to Golden Gate, where both ships leave side by side for China, *Romance* anchors in Hong Kong about an hour first, 251–252; various voyages; at Hong Kong loads for San Francisco and is never heard of afterwards, 253; record passage to California, 376.

Roscoe, New York packet ship, 47.

R. R. Higgins, schooner built for Cape Cod Fleet, 357, 373.

"Sail versus Steam," 104–105; no steamship afloat at the time could have come within 100 miles of *Lightning's* 436 mile log for 24 hours, 270.

St. George, New York packet ship (Red Cross Line), 18, 365.

Salt Pickling (See *Anglo-Saxon*).

Sampson & Tappan, Boston merchants and shipowners, 118, 133, 164, 204, 368, 369; letter from Capt. Richardson of *Stag Hound*, 128; facsimile of letter to Donald McKay, 205.

Sampson, Capt. Gaius, *Daniel Webster*, 106; lost overboard, 96.

Samuels, Capt. Samuel, *Dreadnaught*, 18.

Santa Claus, medium clipper ship, 300, 371.

Savannah, pioneer steam vessel, 33.

Scene in a New York shipping office at the height of the California gold fever, 167–170.

Schley, Commander Winfield S., *Essex*, 342, 372.

Schomberg, British clipper, belonging to Baines' Australian Black Ball Line, 266.

Scientific American, New York magazine, letter from Donald McKay referring to filling in concave lines of his ship *Lightning* by her English owners; also stating he saw recorded in that ship's log (of 24 hours) 436 nautical miles, 266–267.

Seamen's Wages, List of (1848), 59.

Sears, E. & R. W., Boston merchants and shipowners, 133.

Sears, J. Henry, Boston merchant and shipowner, 324.

Seaver, Hon. Benj., Mayor of Boston, 303; speech at launching of ship *Defender*, 308.

391

INDEX

Star of Empire, Boston–Liverpool packet (Train's White Diamond Line), 369; her figurehead and ornaments on stern, 103; construction details, 104; small mortality on prolonged Atlantic crossing creates favorable comment, 107; lost at sea, 107.

Steam navigation for ocean voyages long-delayed, 33–35.

Steers, George, designer and builder of yacht *America,* 346.

Stevens, John C., sells yacht *America* in England, 345.

Sultana, barque, Mediterannean trade, 367.

Surprise, California clipper ship, 144, 212.

Swallow Tail Line, New York to Liverpool packets, New York to London packets, 33. (*See also* Grinnell, Minturn & Co.)

Swordfish, California clipper ship, race with *Flying Fish,* 165–167.

Theodore D. Wagner, wooden screw propeller, 374.

Three Brothers (once the *Vanderbilt* of Commodore Vanderbilt's New York to Havre line), 295.

Thwing, William, & Co., Boston shipowners, 373.

Timber Contracts with British Admiralty, etc., 353

Toucey, I., Secretary of U. S. Navy, 353; letter commending Donald McKay to ex-President Pierce, 354.

Trade Wind, California clipper ship, races with *Flying Fish,* 170–176.

Train, Enoch, Boston merchant-owner, 303, 307, 366, 369, 370; meets Donald McKay and contracts for first ship of his "White Diamond Line," Boston–Liverpool packets, 20; invites builder to "come to Boston" and makes arrangements for financing shipbuilding operations, 21, 98; daring hardihood that impelled him to start line of sailing packets, 36–37; orders a craft A1, in construction to attract passengers on both sides of the Atlantic, 42–43; loses this costly ship but keeps right on maintaining splendid packet service, 46; 97–98; culminates contract for *Ocean Monarch,* 64; last ship built for him by Donald McKay, 107, 109; regrets parting with *Flying Cloud,* 142; introduces the Hon. Benjamin Seaver at *Defender* launching luncheon, 308; his speech upon that occasion, 309–310; an incorporator of the Boston & European S. S. Co., 352; expresses utmost friendliness to this American steamship competition and offers at a public meeting to assist the enterprise, 353. (*See also* Enoch Train & Co.)

Train, Enoch, & Co., Boston merchant-owners, 69–70, 78–79, 96, 99, 144–145, 365, 366, 368; contemplated purchasing *Sovereign of the Seas,* 178. (*See also* Enoch Train.)

Train, George Francis, 70–71; moralizing story from "Youthful Speculations," 40–41; vivid account loss of *Ocean Monarch,* extract from his biography, *My Life in Many States,* 65–66; organizes Caldwell, Train & Co., in Melbourne, 80–81; his account of rescue passengers from ship *Unicorn,* 93–95; claims credit for creating name of *Flying Cloud,* 141–142; attributed reason for not naming *Sovereign of the Seas,* "Enoch Train," 178–179; extract from his *American Merchant Letters* portraying shipping events that startled Australia, 277–278.

Train's Line, Boston–Liverpool packets (*See* White Diamond Line, *also* Enoch Train & Co.).

Trans-Atlantic Packet Service, 36–37; record of short passages, 289.

Treadwell, Captain, *Bald Eagle,* 218, 220.

Trefoil, wooden propeller gunboat U. S. Navy, 373; construction details; placed in service, out of commission, sold, 331–332.

Tyler, General John S., Boston, 317, 318–319.

Ulrich, Mayo & Co., Boston shipowners, 373.

Upton, George B., Boston, shipowner, 79, 82, 83, 118, 133, 164, 211, 225, 250, 312, 367, 368, 369, 370, 372, 373; one of incorporators Boston & European S. S. Co., 352.

Urquhart, Capt. W. W. (*Reminiscences of the Merchant Marine*), 50; interesting account of a New York packet departing from her pier, 51.

393